RA
1231
.S45
N37
1976

Medical and Biologic Effects of Environmental Pollutants

SELENIUM

Committee on
Medical and Biologic Effects of
Environmental Pollutants

DIVISION OF MEDICAL SCIENCES
ASSEMBLY OF LIFE SCIENCES
NATIONAL RESEARCH COUNCIL

NATIONAL ACADEMY
OF SCIENCES
WASHINGTON, D.C.   1976

Other volumes in the Medical and Biologic Effects of Environmental Pollutants series (formerly named Biologic Effects of Atmospheric Pollutants):

ASBESTOS (ISBN 0-309-01927-3)
CHROMIUM (ISBN 0-309-02217-7)
FLUORIDES (ISBN 0-309-01922-2)
PARTICULATE POLYCYCLIC ORGANIC MATTER (ISBN 0-309-02027-1)
LEAD (ISBN 0-309-01941-9)
MANGANESE (ISBN 0-309-02143-X)
VANADIUM (ISBN 0-309-02218-5)
NICKEL (ISBN 0-309-02314-9)
VAPOR-PHASE ORGANIC POLLUTANTS (ISBN 0-309-02441-2)

NOTICE: The project that is the subject of this report was approved by the Governing Board of the National Research Council, whose members are drawn from the Councils of the National Academy of Sciences, the National Academy of Engineering, and the Institute of Medicine. The members of the Committee responsible for the report were chosen for their special competences and with regard for appropriate balance.

This report has been reviewed by a group other than the authors according to procedures approved by a Report Review Committee consisting of members of the National Academy of Sciences, the National Academy of Engineering, and the Institute of Medicine.

The work on which this publication is based was performed pursuant to Contract No. 68-02-1226 with the Environmental Protection Agency.

Library of Congress Catalog Card Number 76-40687
International Standard Book Number 0-309-02503-6

Available from
Printing and Publishing Office
National Academy of Sciences
2101 Constitution Avenue
Washington, D.C. 20418

Printed in the United States of America

80 79 78 77 76   10 9 8 7 6 5 4 3 2 1

SUBCOMMITTEE ON SELENIUM

SAMUEL A. GUNN, University of Miami School of Medicine, Miami, Florida, *Chairman*
JAMES R. HARR, Pennwalt Corporation, Rochester, New York
ORVILLE A. LEVANDER, Agricultural Research Center, Beltsville, Maryland
OSCAR E. OLSON, South Dakota State University, Brookings, South Dakota
HAROLD J. SCHROEDER, U.S. Bureau of Mines, Washington, D.C.

W. H. ALLAWAY, U.S. Plant, Soil, and Nutrition Laboratory, Ithaca, New York, *Consultant*
HUBERT W. LAKIN, U.S. Geological Survey, Denver, Colorado, *Consultant*

T. D. BOAZ, JR., Division of Medical Sciences, National Research Council, Washington, D.C., *Staff Officer*

COMMITTEE ON MEDICAL AND BIOLOGIC EFFECTS OF ENVIRONMENTAL POLLUTANTS

HERSCHEL E. GRIFFIN, Graduate School of Public Health, University of Pittsburgh, Pittsburgh, Pennsylvania, *Chairman*

DAVID M. ANDERSON, Industrial Relations Department, Bethlehem Steel Corporation, Bethlehem, Pennsylvania

RICHARD U. BYERRUM, College of Natural Science, Michigan State University, East Lansing

RONALD F. COBURN, University of Pennsylvania School of Medicine, Philadelphia

T. TIMOTHY CROCKER, University of California College of Medicine, Irvine

SHELDON K. FRIEDLANDER, California Institute of Technology, Pasadena

SAMUEL A. GUNN, University of Miami School of Medicine, Miami, Florida

ROBERT I. HENKIN, Georgetown University Hospital, Washington, D.C.

IAN T. T. HIGGINS, School of Public Health, University of Michigan, Ann Arbor

JOE W. HIGHTOWER, Department of Chemical Engineering, Rice University, Houston, Texas

ORVILLE A. LEVANDER, Agricultural Research Center, Beltsville, Maryland

DWIGHT F. METZLER, Kansas State Department of Health and Environment, Topeka

I. HERBERT SCHEINBERG, Albert Einstein College of Medicine, Bronx, New York

RALPH G. SMITH, School of Public Health, University of Michigan, Ann Arbor

GORDON J. STOPPS, Department of Health, Toronto, Ontario, Canada

F. WILLIAM SUNDERMAN, JR., University of Connecticut School of Medicine, Farmington

BENJAMIN L. VAN DUUREN, New York University Medical Center, New York

BERNARD WEISS, University of Rochester Medical Center, Rochester, New York

T. D. BOAZ, JR., Division of Medical Sciences, National Research Council, Washington, D.C., *Executive Director*

Preface

The Panel on Selenium met for the first time on September 27, 1972, with Dr. W. H. Allaway as Chairman. Dr. Allaway resigned on October 22, becoming a consultant to the Panel, and was succeeded as Chairman by Dr. Samuel A. Gunn. In September 1975 the Panel was redesignated the Subcommittee on Selenium.

This report is an in-depth study that attempts to assemble, organize, and interpret present-day information on selenium and its compounds and the effects of these substances on man, animals, and plants. Emphasis is given to the effects of selenium on man, conclusions are drawn from the evaluation of current knowledge on the subject, and recommendations are made for further research.

The objective of this document is to present a balanced, comprehensive survey of selenium in relation to health for the information of the scientific community and the general public and for the guidance of standard-setting and regulatory agencies. The report tells where selenium is found and how it is measured; describes its physical and chemical nature, its biologic effects, its relation to other pollutants, and dose–response relations, where known; and discusses margins of safety.

Statements are supported by references to the scientific literature whenever possible or are based on a consensus of the members of the Subcommittee.

Acknowledgments

This report was written by the Subcommittee on Selenium, each section being prepared by a member of or consultant to the Subcommittee. Some of the sections and parts of sections were later combined as appropriate. The entire document has been approved by all members of the Subcommittee.

Dr. W. H. Allaway was responsible for the chapter on chemistry and the section on agricultural uses. Dr. Oscar E. Olson contributed the chapter on occurrence, to which Mr. Hubert W. Lakin added information on fossil fuels. Mr. Lakin also throughly reviewed the finished draft of the document. Dr. Olson wrote the section on natural cycling, the section on metabolism in plants, and the chapter on sampling and analysis. He collaborated extensively with Dr. James R. Harr in developing the material on selenosis.

Dr. Harr provided the section on nutritional, prophylactic, and therapeutic uses and the section on carcinogenicity and anticarcinogenicity. He prepared the initial draft of the section on selenosis.

Dr. Samuel A. Gunn was responsible for the sections on the reproductive system, the vascular system, and the medical uses of selenium in human beings. Dr. Orville A. Levander contributed the sections on metabolism in animals, metabolism in microorganisms, and the physiologic role of selenium. Mr. Harold J. Schroeder wrote the section on industrial uses in the chapter on industrial and agricultural uses and the discussion of industrial cycling in the chapter on cycling.

An in-depth presentation of the report was made to the Committee on

Medical and Biologic Effects of Environmental Pollutants by Dr. Olson on February 12, 1974, and comments made by members of the Committee at that time were helpful in further development of the report. The Subcommittee is indebted to Dr. Richard U. Byerrum, who served as Associate Editor, and to those anonymous reviewers who contributed their services.

During its deliberations, the Subcommittee received valuable assistance from Dr. Robert J. M. Horton of the Environmental Protection Agency. Much use was made of the National Research Council's Advisory Center on Toxicology, the National Academy of Sciences Library, the National Library of Medicine, the National Agricultural Library, the Library of Congress, and the Air Pollution Technical Information Center. Acknowledgment is also made of the assistance received from the Environmental Studies Board, National Academy of Sciences–National Academy of Engineering, and divisions of the National Research Council.

Contents

1	Chemistry	1
2	Occurrence	9
3	Industrial and Agricultural Uses	28
4	Cycling	41
5	Biologic Effects	51
6	Sampling and Analysis	134
7	Summary and Conclusions	139
8	Recommendations	149
	References	153
	Index	195

1

Chemistry

The features of selenium chemistry that may control the occurrence, chemical form, and movement of this element in rocks, soils, rivers, groundwater, air, plants, and animal or human tissues are reviewed in this chapter. The atomic properties and electronic structure of selenium are given in standard texts in inorganic chemistry and will not be reviewed here. Some of these properties, as given by Rosenfeld and Beath[636] and Crystal,[135] are as follows:

Atomic weight	78.96
Atomic number	34
Covalent radius Å	1.16
Atomic radius Å	1.40
Ionic radius Å	1.98
Electronegativity	2.55
Oxidation states	−2, 0, +2*, +4, +6
Electronic structure	[Ar] $3d^{10}4s^24p^4$

* The +2 state has not been reported in nature.

There are no naturally occurring radioisotopes of selenium. The isotopes produced by neutron activation and their pathways of decay are listed by Rosenfeld and Beath.[636] The isotopes ^{75}Se, ^{77m}Se, and ^{81}Se may be used in the quantitative measurement of selenium by neutron activation procedures.[802] ^{75}Se is used as a tracer in biologic experiments. ^{75}Se selenomethionine is used in human medicine in certain radiologic diagnostic procedures.

The major features of selenium chemistry that affect its movement, toxicity, and deficiency in the environment are associated with changes in its oxidation state and the resulting differences in chemical properties. The relationships among E_h, pH, and some potential forms of selenium in aqueous inorganic systems, such as weathering rock or soil, are shown in Figure 1-1. From this figure it is apparent that selenium in the +6 oxidation state is stable in alkaline oxidizing conditions. Acid and reducing conditions favor the formation of elemental selenium and selenides.

PROPERTIES OF SELENATE–SELENIUM (+6 OXIDATION STATE)

Selenic acid (H_2SeO_4) is a strong acid ($K_1 = 2$).[704] In solubility, most salts of selenic acid are similar to the sulfates of the same metals. Soluble selenates would be expected in alkaline soils or alkaline weathering rocks in dry areas. Selenates added to soil are taken up by plants, and toxic levels of selenium in plant tissue may result.[509] There is little doubt that soluble selenates are the form of selenium responsible for most naturally occurring instances of plants of high selenium content, even though much of the total selenium in the soil may be present in other forms. Even though one would expect selenate to be converted to selenite or elemental selenium in acid environments, this conversion may be very slow.

Gissel-Nielsen and Bisberg[254] report that crops took up over one-half of the selenate added to a soil of pH 5.7. Substantial uptake of added selenium was evident for some months after selenate was added to this soil, indicating a very slow conversion of selenate to less soluble forms of the element.

Because of its stability at alkaline pH, its solubility, and its ready availability to plants, selenate appears to be the most dangerous form of selenium as far as potential environmental pollution is concerned. Fortunately, any appreciable accidental addition by man of selenate to soil, water, or air appears unlikely.

While the extent to which the selenium level of soils is increased by fertilizer additions is unknown, it has been found that phosphate rocks contain as much as 178 ppm of the element.[626] Much of this is lost during superphosphate preparation. In spite of the lack of information, it appears unlikely that fertilizers other than those to which selenium compounds have been added will add enough of the element to soils to correct nutritional deficiencies.

Chemistry

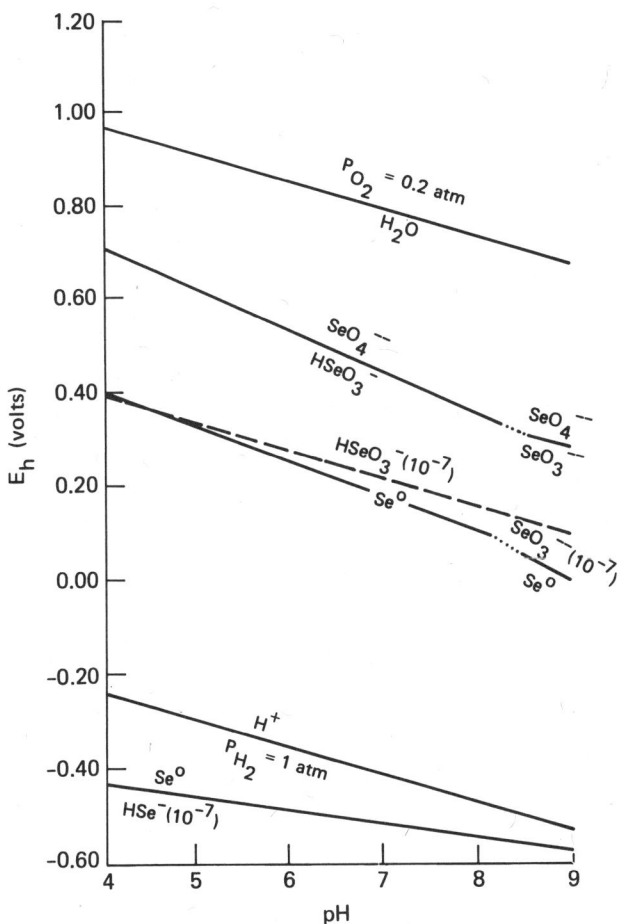

FIGURE 1-1 Relation of oxidation-reduction potentials of some selenium compounds to pH.[246] In the region between the lines for the O_2/H_2O couple and the SeO_4^{--}/SeO_3^{--} couple, selenate would predominate. Between the SeO_4^{--}/SeO_3^{--} and SeO_3^{--}/Se^0 lines, selenite would predominate. Between the SeO_3^{--}/Se^0 and H^+/H_2 lines, elemental selenium and some heavy metal selenides would predominate. At least in soils, hydrogen selenide would not be expected to exist. (The dashed line represents Pearsall's E_h-pH dividing line between oxidized and reduced soils.[598])

PROPERTIES OF SELENITE–SELENIUM (+4 OXIDATION STATE)

Selenious acid (H_2SeO_3) is a weak acid, and any dissolved selenite would be present predominantly as the biselenite ion in waters between pH 3.5 and 9. Most selenite salts are less soluble than the corresponding selenates. Of especial interest with respect to environmental problems is the very low solubility of the ferric selenites. Geering et al.[246] found evidence of a ferric selenite compound or adsorption complex of even lower solubility than any of the known ferric selenites in soils equilibrated with labeled selenite. Cary and Allaway[109] added tagged selenite at the rate of 1 ppm to several soils of low selenium content. Alfalfa grown on these soils in a greenhouse generally contained concentrations of selenium that would not be toxic to animals that ate the alfalfa. Studies of the selenium in these soils several months after addition indicated that most of it, although not water soluble, was isotopically exchangeable with neutral selenite solutions and thus may have been present as a very stable adsorption complex on sesquioxide surfaces of the soil. Added selenite tended to remain more soluble when it was added to a very coarse-textured soil of very low iron content.

Another property of selenite of importance to environmental cycling of selenium is that selenite is rapidly reduced to elemental selenium under acid conditions by mild reducing agents, such as ascorbic acid or SO_2.[636]

The probability that selenite will either form insoluble compounds or adsorbates with ferric oxide or be reduced to insoluble elemental selenium minimizes the hazard of pollution of the environment by inadvertent additions of selenite selenium.

PROPERTIES OF ELEMENTAL SELENIUM (0 OXIDATION STATE)

Different allotropic forms of elemental selenium are listed and their solubility in different reagents is tabulated by Rosenfeld and Beath.[636] The electronic and photoelectric properties of "metallic" selenium are the basis for many industrial uses of this element. As far as environmental problems are concerned, the extreme insolubility of elemental selenium in aqueous systems and the fact that elemental selenium is formed by high-temperature decomposition of most natural materials, such as fossil fuels and organic refuse, are of primary importance.

The stability of elemental selenium is demonstrated by its occurrence in sandstones in dry, alkaline environments.[417] Cary and Allaway[109] added freshly precipitated elemental selenium at the rate of 1 ppm to several different soils and cropped them to alfalfa in a greenhouse. Concentrations of selenium in the alfalfa growing on these soils were well below limits that might be toxic to animals, and after a few months these concentrations in the alfalfa declined to levels that would not have protected animals from selenium-responsive diseases. Handreck and Godwin[310] placed heavy pellets containing elemental selenium in the rumen of sheep without causing any evidence of selenium toxicity.

Elemental selenium burns in air to form selenium dioxide, SeO_2. In the combustion of fossil fuels or organic materials, the SeO_2 formed will be reduced to elemental selenium by the sulfur dioxide that is always formed during combustion of these materials in concentrations greatly in excess of the amount required for reduction of the SeO_2 formed.[810]

It appears that elemental selenium is a major inert "sink" for selenium introduced into the environment in various ways, and contamination of water, soil, or air by elemental selenium poses a minimal hazard of selenium toxicity. At the same time, fly ash from the combustion of fossil fuels may contain sufficient elemental selenium to represent a major waste of an expensive and scarce natural resource.

PROPERTIES OF SELENIDE-SELENIUM (−2 OXIDATION STATE)

Hydrogen selenide is a fairly strong acid, and its fumes are very toxic. However, this compound rapidly decomposes in air to form elemental selenium and water; thus, hazard from hydrogen selenide is confined to industrial installations.

The selenides of heavy metals are very insoluble. For mercuric selenide the K_{so} is given as −59,[704] and the formation of insoluble mercuric selenide may be a major mechanism involved in the detoxification of methyl mercury by dietary selenite.[240] Other selenides, such as those of copper and cadmium, are also of low solubility.

It appears that considerable amounts of insoluble selenides, or possibly elemental selenium, are contained in the feces of ruminant animals that consume dietary selenium.[606] It is impossible to differentiate heavy-metal selenide selenium from elemental selenium in fecal material or similar organics by chemical means. The selenium present in fecal material apparently is not readily taken up by plants when the fecal material is applied to soil.

Selenium is present in many pyrites and sulfide ores.[324,825] There are no records of highly seleniferous plants growing near exposures of pyrites or sulfide ores in humid regions where acid or neutral soils occur over the pyritic deposits. Thus, metal selenides, as well as elemental selenium, may represent a useful inert "sink" for detoxification of selenium added to these areas.

In semiarid and arid regions, selenide-selenium appears to have been oxidized, over geologic time, to selenate. Plants containing toxic levels of selenium may be found where seleniferous rocks have weathered in these regions, limited leaching having permitted the accumulation of soluble selenate.

BIOCHEMISTRY OF SELENIUM

Details of the forms of selenium in plants and animals are presented in the sections on selenium metabolism. In this section, only features of the biochemistry of selenium that may affect its tendency to recycle in biologic systems or its potential hazard as an environmental pollutant will be discussed. For a review of the biochemistry of this element from the standpoint of its role in enzyme systems, the reader is referred to a recent article by Stadtman.[725]

Selenite and selenate are both taken up by the roots of plants. Within the plant these forms of selenium are reduced to the -2 oxidation state, and the Se^{-2} is incorporated into soluble amino acids, protein-bound amino acids, or both. The reduction of selenium within the plant may not be quantitative; this is especially true for selenates that are taken up by the roots.

Monogastric animals may reduce selenate and selenite, but they apparently do not incorporate the reduced selenium into amino acids. The "selenotrisulfides" formed by the reaction of selenite with sulfhydryl groups of amino acids, peptides, and proteins are probably the first products of the reduction of physiologic doses of selenite in monogastric animals.[235] Major excretory products of selenium metabolism in animals are trimethyl selenonium ion in the urine[585] and elemental selenium or metal selenides in the feces.[606]

Some of the chemical and biochemical changes possibly involved in the movement of selenium from soils through plants and animals are diagrammed in Figure 1-2. When the metabolic pathways of selenium in plants and animals are considered along with the reactions of selenium in soils, it appears that conversion of the element to inert and insoluble forms is a feature of the soil–plant–animal system. Where such a system is confined to an area of acid or neutral soils, and no selenium is added

Chemistry

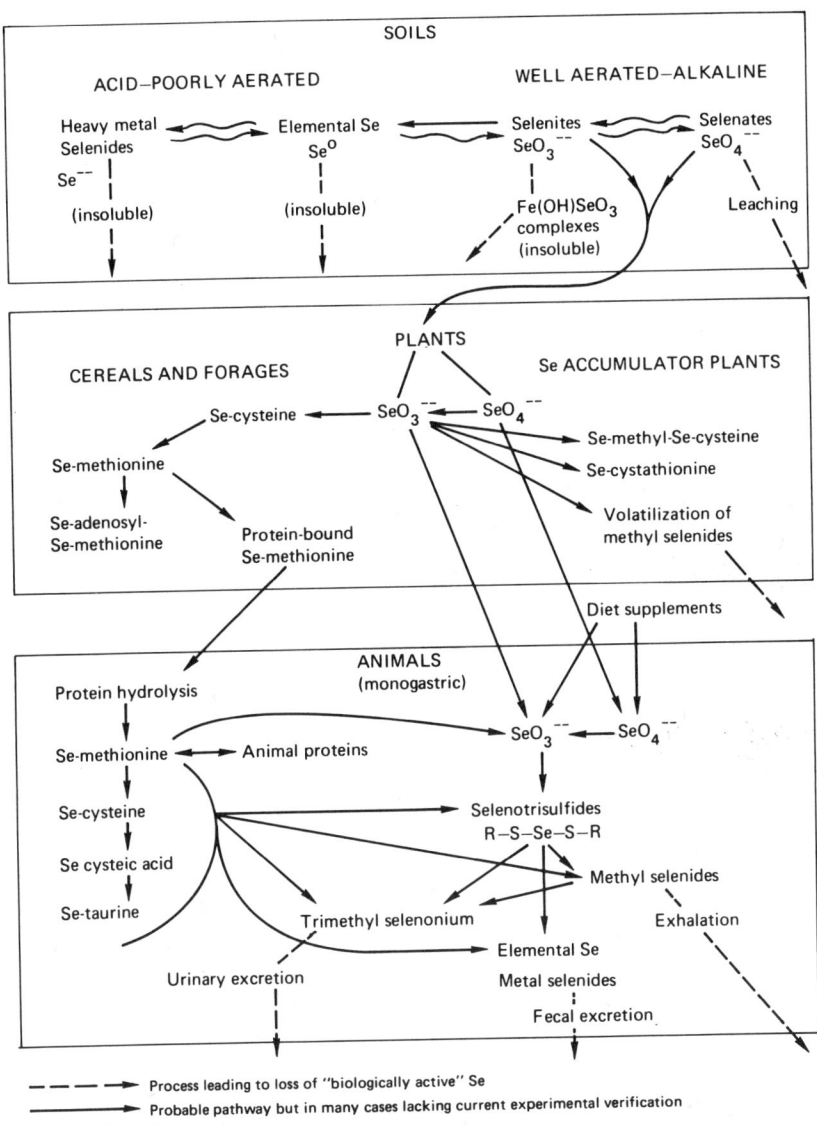

FIGURE 1-2 Chemical and biochemical changes in selenium possibly involved in its movement from soil through plants to animals.[8]

to the system, the amount of "biologically active" selenium should steadily decline. If the soil parent material contains some selenium, as for example in western Iowa, the time required to lower selenium concentrations in animal diets to deficiency levels may be thousands of years. Where a soil–plant–animal system operates in an arid area of alkaline soils, the selenium returned to the soil in plant residues or animal excreta may be reoxidized to selenate rapidly enough to maintain the level of biologically active selenium at a nearly constant level.

ORGANIC CHEMISTRY OF SELENIUM

Important features of the chemistry of selenium in synthetic organic compounds were reviewed recently in the proceedings of a symposium edited by Okamoto and Günther[557] and by others.[396] As far as problems of environmental contamination are concerned, the important feature of the synthetic organic chemistry of selenium is that essentially all of the compounds synthetized contain the selenium in the −2 oxidation state. These compounds might be expected to decompose to form elemental selenium; in fact, the tendency of some of these compounds to decompose and form elemental selenium is one of the major problems encountered in working with them. Since elemental selenium is inert and generally nontoxic, the main hazard of selenium toxicity to people involved in the synthesis of organic selenium compounds is from the compounds themselves or from intermediates in their preparation.

2

Occurrence

GEOLOGIC

The geology and geochemistry of selenium have been reviewed by a number of authors.[18,416,509,641] Except where indicated otherwise, the following discussion stems from these reviews.

Concentration in Earth's Crust

Although many geologic specimens have been reported to contain no selenium, it is probable that with sufficiently sensitive methods the element can be found in all rocks and soils. Most estimates of its average concentrations in the earth's crust range from 0.03 to 0.8 ppm, but several fall around 0.1 ppm. It is usually found at concentrations of less than 1 ppm, except in soils or parent materials where selenium poisoning is a problem, in some mineral deposits, and in certain acid or ferruginous soils.

During the cooling and crystallization of magmas, their selenium content may be diminished, either by volatilization or because of the element's tendency to remain with the liquid portion and to flow into fractures or dissolve into adjacent rocks. Nevertheless, igneous rocks have been estimated to contain an average of about 0.09 ppm of the element, and, because they constitute so large a portion of the earth's crust, this value is accepted by some as the average for crustal abundance. In most cases, igneous rocks would not be expected to contain over 1 ppm of selenium.

Occurrence in Sulfides, Sulfates, and Sulfur

Chemically, selenium resembles sulfur, and sulfide or native sulfur deposits often contain it in significant amounts. Thus, sulfides of bismuth, iron, mercury, silver, copper, lead, and zinc have been found to contain the element, occasionally at levels of over 20%.[144,453] Jarosite and barite, two sulfate minerals, have also been found to contain selenium, but at relatively low levels. Crude sulfur also often contains selenium, sometimes well over 0.1%. Deposits of sulfur-containing minerals are often secondary in nature, and when selenium occurs in them it probably has been leached from some other material and redeposited. It also appears to have crystallized as metallic selenides associated with sulfides in epithermal rocks.[144]

Sandstone

Although sandstones have been found to contain highly variable amounts of selenium, many probably contain less than 1 ppm.[503,509] However, because sandstones are somewhat porous, waters may enter them from adjacent formations. The waters may carry and deposit selenium, usually with iron minerals. Sandstones containing over 100 ppm of the element have been reported in Wyoming.[43,403]

Limestone

Although the selenium content of limestones is usually very low, values of over 40 ppm have been reported in the chalky shales and marls of the Niobrara formation of South Dakota. Phosphate rocks range from well below 1 ppm to about 300 ppm, suggesting that phosphate fertilizers may, on rare occasions, provide significant amounts of the element to soils deficient in it. Limonitic concretions and meteoritic materials have been found to contain selenium in amounts from less than 1 to over 200 ppm.[830]

Other Sedimentary Rocks

Of the sedimentary rocks, shales seem consistently higher in selenium than limestones or sandstones, and in the United States the soils derived from them are, in the main, responsible for the problem of selenium poisoning in animals. Although their selenium content varies both vertically and areally over a wide range, shales provide enough consistency to make a knowledge of the geology of a region of great assistance in locating soils

Occurrence

of excessive selenium content. For instance, the Mobridge member of the Pierre formation near the Missouri River in southern South Dakota, although it ranges from less than 1 to over 30 ppm in its selenium content throughout its profile, is generally highly seleniferous, and where it outcrops it has the potential to form soils capable of producing toxic vegetation. Only 100 miles north or east, its selenium content has markedly decreased, and it no longer weathers to seleniferous soils. On the other hand, the Smoky Hill member of the Niobrara formation is generally highly seleniferous wherever it outcrops in the state. This type of information greatly aids in the mapping of potentially toxic areas.

Geologic History

In the United States, the most highly seleniferous sediments were laid down in the shallow seas of the Cretaceous period, but the origin of selenium in these sediments has not been definitely determined. Byers et al.[97] found rather high levels of selenium in some ferruginous soils of Hawaii, especially in areas of high rainfall. Finding selenium in the volcanic gas of the area, they concluded that the selenium in these soils was derived chiefly from these gases and associated sublimates carried down by rains and fixed by the soils in a highly insoluble form. They extended this conclusion to sedimentary rocks of the continental United States, advancing the following as supportive evidence: (1) bentonite deposits (presumed volcanic in origin) precede, accompany, and follow selenium deposition in the Cretaceous period of geologic history; and (2) selenites are absorbed and precipitated by iron oxides and may thus have been removed from the seawaters into which the rains had fallen and concentrated in the sediments. Volcanic tuffs[43] and volcanic sulfur[97,737] of very high selenium content have been reported, lending credence to the volcanic origin theory, and the data of Davidson and Powers,[145] who found the selenium content of crystalline (slow-cooling) volcanic rocks lower than the content of those not crystalline (fast-cooling), support the theory. Further, Howard[350] suggests that the higher selenium content of the Smoky Hill (Niobrara) and lower Pierre shales indicates a volcanic origin for the element. He concludes from thermodynamic calculations that H_2Se and Se^{-2} are the expected forms in magmatic gases; Se^{-2}, Se^{+6}, and H_2Se could exist in fumarolic and vent gases; SeO_2 could form on eruption into an oxygen-rich atmosphere; and, on cooling, both oxidized and elemental selenium will condense and be deposited with, but not as an integral part of, the volcanic ash.

Coleman and Delevaux[126] analyzed many sulfide mineral samples from the western United States. Their data suggest volcanic activity or hydrothermal fluid extraction from magmatic sources and other seleniferous beds as the source of selenium in sedimentary rocks. They state that on the Colorado Plateau and in Wyoming the selenium in sulfides can be related to a magmatic province that was very high in selenium during periods of volcanic and extrusive activity in Tertiary and Mesozoic times. In view of this, and since some very seleniferous shales show no evidence of volcanic activity during or shortly before the time they were laid down,[42] the primary origin of at least a part of the selenium in sedimentary rocks may have been igneous or other sedimentary rocks from which the element was leached. In regions where selenium excesses are not a problem, volcanism was probably not an important source of the element.

The selenium in some soils may well have had its origin in other soils or sedimentary rocks lying at higher elevations. Thus, the highly seleniferous soils of Ireland have apparently resulted from the transport by water of soluble forms of the element from a sedimentary rock formation into a poorly drained basin containing much organic matter where reduction and precipitation of the element occurred.[204,798,806] In Israel, an alluvial soil apparently derived its selenium from a higher-lying limestone.[620]

Because of the apparent role of volcanic activity in the development of seleniferous geologic beds, some have suggested bentonite as a source of selenium for feeds deficient in the element. Selenium has, indeed, been found to occur at fairly high levels in some bentonites. It was not indicated that the samples taken at or near an exposed surface were free from secondary deposits of the element or from contamination by adjacent shales. Examination of several South Dakota bentonites has suggested that they would not be a reliable source of the element for the purpose of feed supplementation.[562]

Selenium from Rocks to Soils

Selenium probably occurs as the free element or, more likely, as a metal selenide in unweathered rocks. It is apparently readily oxidized during the weathering of parent materials to soils. In areas of acid soils, the element would probably be present as the selenite firmly bound in iron oxide colloids, while in alkaline soils it would oxidize further to the very soluble selenate. It has been suggested that the primary accumulator plants (see the discussion of plant metabolism in Chapter 5) are capable of converting insoluble and thus unavailable selenium to a soluble form, but the evidence for this is not convincing. It would appear that the normal weathering processes, probably including microbial activity, could account for the

Occurrence

conversion of the unavailable form of the element in soils to the available forms.

In highly seleniferous soils, it seems that the available form of the element is represented largely or almost entirely by selenate.[573] The data of Krauskopf on the oxidation potentials of selenium have been discussed in terms of their meaning in the weathering of soils,[18] and under alkaline conditions the oxidation of the element takes place with relative ease. In his discussion of the selenium cycle in nature, Shrift[689] points to the scant information concerning the biologic oxidation of the element. It is tempting, therefore, to accept chemical weathering as almost solely responsible for the oxidation of selenium during soil formation. However, there are data suggesting that biologic oxidation is also important,[246] and this matter needs more attention.

The selenium content of soils depends on many factors. The most important of these seem to be the selenium content of the parent materials and the intensity of weathering and leaching. A number of conditions influence the availability of the element for absorption by plants, and these are discussed in Chapter 5. This available selenium can be evaluated by plant analysis,[374] and Kubota *et al.*[409] have applied this concept in preparing the map shown in Figure 2-1. The map also suggests in a general way the areas of soils of high and low total selenium content.

FOOD CHAIN

Figure 2-1 gives a general picture of where in the United States feedstuffs used for animal production might contain deficient, optimal, or excessive levels of selenium. Normally, of course, animal diets contain a variety of supplements, and some of these contribute significant amounts of the element. Also, in the case of poultry and swine, specific selenium supplements are being used. Although the data in several reports suggest that feeds highest in protein are generally highest in selenium, Scott and Thompson[672] conclude that for plant materials the selenium values depend largely on the level of the element in the soils where the plants were grown.

Foods with High Selenium Content

SELENIFEROUS AREAS

In 1935, Byers[93] reported the selenium content of some foods produced on seleniferous farms in South Dakota, finding up to 1.2 ppm in whole

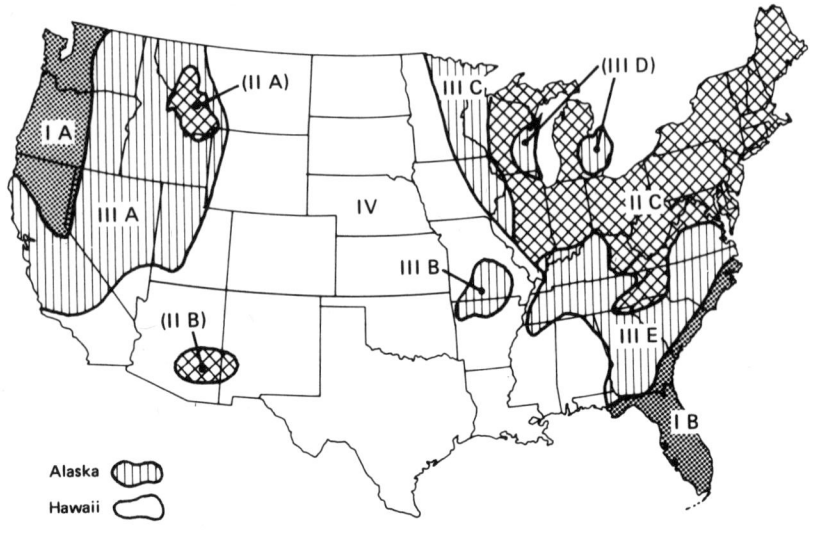

FIGURE 2-1 Selenium in crops in different regions of the United States.[409] *Data from USDA Technical Bulletin 753, 1941.

milk; 10 ppm in whole egg; and 2, 7, 25, and 100 ppm (presumably on a dry-matter basis) in string beans, lettuce, turnip leaves, and cabbage, respectively. It should be stressed here that these values are unusually high, since they represent the exceptional case of a highly seleniferous area. The average selenium concentration in the diets of families in such a seleniferous area would have been considerably lower than these data suggest. Because of our present-day food distribution, the concentration

Occurrence

would now be even lower. The same is true for the reports referred to below.

In 1937, Smith and Westfall[714] reported analyses as follows for locally produced foods from the same general area that Byers had sampled: milk—of 50 samples, selenium was detected in 44 at levels up to 1.27 ppm; eggs—selenium was detected in all of 32 samples at levels between 0.25 and 9.14 ppm; meat (muscle)—selenium was detected in all of 6 samples at levels between 1.17 and 8 ppm; bread (made from locally milled flour)—selenium was detected in all of 11 samples at levels up to 1 ppm; vegetables—selenium was detected in all but 4 of 99 samples at levels up to 17.8 ppm. Williams et al.[830] reported 0.6 ppm of selenium in milk from a Mexican ranch and up to 70 ppm (apparently on a dry-matter basis) for a number of vegetables from a Mexican market in a seleniferous area. On the other hand, they found 3 ppm or less (dry-matter basis) in vegetables raised in an irrigated garden on a seleniferous South Dakota ranch. Later studies suggest that the reason for these low values was that the garden was situated on a part of the ranch that did not produce highly seleniferous plants.[156]

WHEAT AND WHEAT PRODUCTS

Robinson[630] analyzed samples of market wheat from various parts of the world, finding levels between 0.1 and 1.9 ppm of selenium. Thorvaldson and Johnson,[751] following a report of high concentrations of the element in young wheat plants in Saskatchewan and Alberta,[95] analyzed 230 composites made up from 2,230 samples of wheat from the former province, finding an average value of 0.44 ppm with a maximum of 1.5 ppm. On analyzing 951 samples of wheat from eight states in the more seleniferous region of the United States, Lakin and Byers[414] concluded that, in view of their findings and those of Robinson,[630] selenium could probably be detected in all wheat. They found that 82.5% of their samples contained 1 ppm or less and 7.5% contained over 4 ppm. Of 66 samples of flour milled in the area, only 5 contained more than 1 ppm, the maximum value being 5 ppm. Similar concentrations were found in bran, shorts, and middlings, which is in agreement with a study indicating that, for wheat containing over 1 ppm, the element distributes itself quite uniformly in the various mill fractions.[510] It should be pointed out, however, that selenium concentrates in the gluten fraction four to five times more than it does in the whole wheat.[570,630] Although other grains used for foods have not been as intensively studied, available data are similar to those for wheat.[93,503,592,830,831] Recently it has been reported that rather small losses of selenium occur during the manufacture of breakfast cereals from grains, and what is lost appears in the by-products destined for animal foods.[195,196]

MUSTARD SEED, BEANS, SUGAR

Williams et al.[831] examined 23 samples of mustard seed, finding 5 ppm of selenium in 1 and 3 ppm or less in the others. Of 17 samples of dry beans, 1 contained 3 ppm; 1, 2 ppm; and the rest, 1 ppm or less. A sample of sugar produced in a South Dakota plant contained less than 0.1 ppm.

Foods from Nonseleniferous Areas

Most of the reports cited above were by early investigators concerned with the toxicity of selenium, and they therefore represent, to a large extent, foods from our more seleniferous areas or foods of the types most likely to contain higher levels of the element. More recent data discussed below are, in general, more representative of our normal diet.

EGGS AND MILK

Hadjimarkos and Bonhorst[293] reported that in Oregon the selenium content of 73 egg samples averaged 0.317 ppm and that of 67 milk samples averaged 0.034 ppm. They found that most of the element in the eggs was concentrated in the yolk, an observation previously reported by Taussky et al.[745] Later, Hadjimarkos[768] reported a mean of 0.021 and a range of 0.013–0.062 ppm for 15 samples of human milk in Portland, Oregon.

Fink[201] found values of 0.076–0.374 ppm of selenium for dried milks from various sources. Considerable loss of the element occurred during preconcentrating prior to drying.[202] Sixty samples of German milks that were freeze-dried contained 0.05–0.13 ppm of selenium, 27 samples of milk powder contained 0.088–0.152 ppm, and drum-drying was reported to cause a loss of only 4.3%–4.7% of the element.[390] Milk from 10 cows in Denmark was found to contain an average of 0.2 ppm on a dry basis.[54] Allaway et al.[12] reported that a sample of whole milk from Rapid City, South Dakota, which lies in an area where selenium is often in excess in the soils, contained 0.05 ppm and that a sample from Bend, Oregon, where selenium deficiencies are known to exist, contained 0.02 ppm. These data suggest that the selenium contents of milks from various areas may vary

Occurrence

over a relatively small range only and that whole milk seldom should contain over 0.05 ppm of the element.

FISH

The selenium content of various fishmeals has been found to vary within and between types, the values falling between 0.15 and 6.70 ppm.[391,392] The overall average appeared to be about 2 ppm. The selenium in fish and shellfish in Japanese waters was found to range from 0.05 to 3.64 ppm.[735] Of 438 fish of a variety of species taken from New York waters, almost all had a selenium content of less than 1 ppm on a wet basis, and concentration of the element did not appear to increase with the age of the fish.[583] These reports and others[499,556] suggest that fish, at least those taken from the ocean, may be a generally good dietary source of the element. However, the nutritional availability of selenium in fish products may be low.[101]

MEATS

Ku et al.[406] reported on the selenium content of longissimus muscle (loin) of swine raised at 13 state experiment stations on diets typical in the areas. Average values (wet basis) for these states ranged from 0.034 ppm (Virginia) to 0.521 ppm (South Dakota). Data for a variety of meats prepared in South Dakota are shown in Table 2-1. With few, if any, ex-

TABLE 2-1 Selenium Content of Some Meats Processed in South Dakota[a]

Kind of Meat	No. of Analyses	Average Selenium Content (ppm—wet basis)
Liver		
Beef	5	0.58
Pork	4	0.70
Chicken	3	0.80
Muscle		
Beef	5	0.21
Pork	5	0.31
Lamb	5	0.32
Chicken	6	0.42
Turkey	12	0.46
Wild pheasant	55	0.51
Processed meats	55	0.33

[a] O. E. Olson, unpublished data.

TABLE 2-2 Selenium Values for Poultry Tissues from Birds Fed Two Practical-Type Diets[672]

Selenium in Diet (ppm)	Selenium Content (ppm—wet basis)					
	Chicks			Poults		
	Muscle	Liver	Skin	Muscle	Liver	Skin
0.07	0.06	0.25	0.09	0.06	0.15	0.07
0.67	0.29	0.80	0.25	0.32	1.03	0.36

ceptions, these meats were from animals taken or produced where neither selenium poisoning nor deficiency is a problem. Where selenium poisoning is a problem, values of 5.6 ppm for liver and 3 ppm for muscle have been reported in experimental animals.[514] However, the likelihood of such meats reaching markets is extremely small, since animals are not usually finished for marketing on toxic feeds. Game animals are not likely to accumulate high levels of the element, because they roam over wide areas in their search for food.

POULTRY

Scott and Thompson[672] studied selenium values for poultry tissues from birds that were fed two practical-type diets (Table 2-2). The differences in the selenium content of these diets resulted from the inclusion of soybean oil meals of low or high selenium content.

COFFEE AND TEA

Shah et al.[677] reported on the trace element content of coffee and tea. By analysis with nondestructive neutron activation, they found the following: ground coffee, concentrations of selenium ranging from 0.028 to 0.204 ppm and averaging 0.124 ppm for five samples; instant coffee, concentrations ranging from 0.004 to 0.170 ppm and averaging 0.069 ppm for four samples; and a concentration of 0.116 ppm for one sample of tea.

GENERAL DIETARY LEVELS IN DIFFERENT COUNTRIES

In Japan, the selenium contents of meat, eggs, and cereals were found to be 0.01–0.05, 0.12–0.26, and 0.02–0.87 ppm, respectively.[735] Foods from 22 Australian villages were found to contain 0.01–0.14 ppm.[176] The mean selenium contents of some foods from the Ukrainian Soviet Socialist Republic were as follows:[733] cabbage, 0.063 ppm; wheat bread, 0.280 ppm;

rye bread, 0.275 ppm; peas, 0.281 ppm; potatoes, 0.142 ppm; onions, 0.096 ppm; beets, 0.139 ppm; carrots, 0.093 ppm; cucumbers, 0.058 ppm; apples, 0.004 ppm; meat, 0.292 ppm; milk, 0.100 ppm; eggs, 0.022 ppm; and cottage cheese, 0.298 ppm. The author concluded that chronic selenium intoxication in man or livestock should not be expected in this region, but that the possibility of a deficiency of the element could not be excluded.

Analysis of a number of vegetable, milk, and egg samples from various parts of Venezuela has established two areas in that country that produce highly seleniferous food.[494] Many vegetable samples contained over 3 ppm of the element. Milk samples ranged from 0.05 to 0.206 ppm; eggs, from 0.49 to 2.34 ppm. Later studies suggest that even in the more seleniferous areas levels in foods do not pose a serious health hazard.[369]

The selenium contents of a number of Egyptian foods have been reported;[466] almost all were below 0.4 ppm on a dry-weight basis, and most were below 0.05 ppm. Differences in methods of drying caused large differences in the values, and this should be taken into account in evaluating the results.

Oelschläger and Menke[556] reported the selenium content of many German foods. Their results are summarized in Table 2-3. These values, when converted to a wet basis, agree in general with those of Morris and Levander,[499] which are discussed below.

The recent data of Morris and Levander[499] for a cross section of the American diet are particularly helpful in assessing selenium intake by people in the United States. They are summarized in Table 2-4. The rather high level of selenium in garlic is confirmed by other data.[265] Higgs et al. point out that cooking may cause selenium losses in some cases, but for most foods the losses are minor.[337]

Hopkins and Majaj[344] stated that when five total human diet samples were collected in the Baltimore, Maryland, area at quarterly intervals

TABLE 2-3 Selenium Content of Certain German Foods[a]

Food	Average Selenium Content (ppm—dry basis)
Meats (pork and beef muscle)	0.27
Liver (pork and beef)	0.44
Milk (dried)	0.14
Fish (ocean)	1.54
Eggs (dried whole)	1.01
Vegetables and fruits	0.03–0.30 (range)

[a] Derived from Oelschläger and Menke.[556]

TABLE 2-4 Selenium Content of Certain Foods in the American Diet[a]

Food	Average Selenium Content (ppm—wet basis)[b]
Vegetables (canned and fresh)	0.010 (0.004–0.039)
Fresh garlic	0.249
Mushrooms (canned and fresh)	0.118
Fruits (canned and fresh)	0.006 (<0.002–0.013)
Cereal products	0.38 (0.026–0.665)
Egg whites	0.051
Egg yolks	0.183
Brown sugar	0.11
White sugar	0.003
Cheeses	0.082 (0.052–0.105)
Table cream	0.006
Whole milk	0.012
Meats (excluding kidney)	0.224 (0.116–0.432)
Seafoods	0.532 (0.337–0.658)

[a] Derived from Morris and Levander.[499]
[b] Figures in parentheses indicate range in values.

during 1963 and 1964 and analyzed for selenium by neutron-activation analysis, none of the element could be detected. The authors point out, however, that these limited results do not indicate a deficiency of the element in the American diet.

An interesting review of the movement of selenium in the food chain has been prepared by Allaway,[8] who concludes that any soil–plant–animal chain of food production that is operating on acid or neutral soils will ultimately become depleted of biologically active selenium. Further, even on alkaline soils there is little, if any, evidence for a significant increase in selenium up the food chain.

The data discussed above suggest that there are wide differences in the selenium content of foods and that they are due mainly to the type of food and where it was produced. However, food eaten in the United States is generally varied in nature and origin, and there seems no reason to expect either inadequacy or excess of the element in our diets except, possibly, under very unusual circumstances.

FOSSIL FUELS

Selenium dispersed by volcanism and by weathering of sulfide deposits is reconcentrated by biogeochemical processes and is enriched in plant

Occurrence

and animal tissues. The enrichment in biomaterials is suggested by the presence of the element in coal deposits. Geologic processes in which decomposition of organic matter occurs are involved in coal formation, and this type of concentration mechanism should not be used to illustrate a food chain buildup. The 138 samples of coal from U.S. deposits reported in Tables 2-5 and 2-6 contained an average of 2.8 ppm of selenium, which is over 25 times the crustal abundance of the element.

Pillay et al.[608] estimated that the annual release of selenium from the combustion of coal and oil in the United States is about 8 million lb. This figure is nearly six times the 1964 production of the element in the whole of North America and four times the world production for the same year.[422] One might expect from these data that the industrial part of the United States would have soils containing an excess of the element. The fact is, however, that 65% of the forage crops in the industrial eastern part of the country (area II C in Figure 2-1) appear to contain insufficient selenium for the growth of healthy animals. There are a number of explanations for this apparent disparity. Selenium fallout as the element

TABLE 2-5 Selenium Content of U.S. Coals[608]

State	Number of Counties Sampled	Number of Samples	Selenium (ppm) Low	High	Average
Alabama	3	4	2.20	8.15	5.14
Colorado	2	3	1.25	2.05	1.65
Illinois	2	2	1.05	1.97	1.51
Indiana	3	4	1.41	8.36	3.96
Iowa	1	1	1.54	1.54	1.54
Kansas	1	1	2.27	2.27	2.27
Kentucky	4	5	1.71	4.82	3.13
Maryland	1	1	1.70	1.70	1.70
Missouri	1	2	3.41	4.98	4.19
Montana	3	3	2.20	4.11	3.04
New Mexico	2	2	4.43	4.82	4.62
North Dakota	1	1	0.98	0.98	0.98
Ohio	4	4	2.64	7.30	4.62
Pennsylvania	7	11	1.35	10.65	3.74
Tennessee	1	1	4.89	4.89	4.89
Utah	2	4	1.30	2.37	1.83
Virginia	3	4	2.24	6.13	4.37
Washington	1	2	0.46	0.66	0.56
West Virginia	12	30	0.92	6.80	3.36
Wyoming	1	1	3.43	3.43	3.43
TOTAL	55	86	—	—	3.36

TABLE 2-6 Selenium Contents of Coals at Active or Proposed Power Plants in the Western Part of the United States[738]

Location of Plants	Average Selenium Content of Coal (ppm)[a]
Four Corners, New Mexico	2.0 (21)
Cholla, Arizona	2.3 (4)
Mohave, Nevada	1.6 (2)
Hayden, Colorado	1.2 (3)
Naughton, Wyoming	0.7 (5)
San Juan, New Mexico	2.2 (2)
Navajo, Arizona	1.6 (2)
Kaiparowits, Utah	1.7 (3)
Huntington Canyon, Utah	1.7 (9)
Jim Bridger, Wyoming	1.5 (1)

[a] Number of samples represented in each average are shown in parentheses.

would not be available to plants and would not be expected to oxidize to an available form in this area of acid soils. Selenium in the form of selenium dioxide would be firmly bound to certain soil colloids and thus not available. Finally, the data of Davis[149] and the section "Industrial" in Chapter 4 suggest that the amount of atmospheric selenium actually derived from the burning of coal is much less than 8 million lb.

Coal

Not all of the selenium released by the burning of coal enters the atmosphere. The coals supplied to five power plants in the western United States were analyzed for selenium and ash content.[738] The bottom ash and fly ash were collected and analyzed, and the approximate ratio of the two types of ash was estimated. It was subsequently estimated that 25,810 lb of the 69,000 lb of selenium in the coal was lost to the atmosphere. In a study in the Denver area, J. W. Kaakinen (personal communication) found the following:

The preliminary results of an experimental mass balance of several trace metals in a coal-fired power plant include some information on the fate of selenium contained in the coal. Selenium concentrations in ashes collected at seven points in the lower power plant were determined by X-ray fluorescence. The measured selenium concentrations ranged from about one part per million in mechanical collector ash to a few hundred parts per million in the fly ash, leaving a wet scrubber. There was a tendency for increasing selenium concentrations in fly ash

Occurrence

samples obtained at successive points downstream from the furnace towards the stack outlet. It may also be noted that as the flue gas proceeds downstream from the furnace its temperature decreases and the average particle size of uncollected fly ash decreases. Analytical determinations of selenium in the raw coal feed and in stack gas vapor have not been completed to date. However, mass balance calculations assuming a conservative figure of 0.5 ppm selenium in the raw feed coal result in less than half of the selenium input accounted for in all the ash and water outputs from the plant, the remainder probably leaving the stack as vapor.

If the selenium content of the coal used in this example had been more in line with the data from Colorado in Table 2-5 (about 2 ppm), over 85% would have left the stack.

Oil

Data on the selenium content of fuel oils are very limited. Bertine and Goldberg[47] estimated the average as 0.17 ppm. Analysis of 47 samples of crude or fuel oils from various parts of the world by neutron activation gave values of from less than 0.006 to 2.2 $\mu g/g$, the average being something less than 0.6 $\mu g/g$.* Hashimoto et al.[322] reported values of 0.50–0.95 (average 0.82) μg of selenium per gram in five samples of raw petroleum and 0.50–1.65 (average 0.99) μg of selenium per gram in nine samples of heavy petroleum. It appears that, on the average, oils contain less of the element than do coals, but more data on both of these fuels are needed.

WATER

The concentration of water-soluble selenium in some soils,[36,38,40,94,96,573] in certain salt crusts or deposits,[96,826] or occasionally in other geologic materials[43] has been documented rather well. Thus, groundwater and surface water should be expected to contain the element, particularly in areas where it is in excess. Indeed, with analytical methods sensitive enough, it might be found in any natural water.

Surface Waters, Drinking Supply

Unfortunately, data on waters are limited. However, those that are available suggest that one would rarely find surface waters containing toxic levels of the element, or even amounts that would contribute sig-

---

*Data on file at Monitoring and Analysis Division, Office of Air and Water Programs, Environmental Protection Agency.

nificantly toward supplying the nutritive requirements of animals.[535] For instance, Lakin and Davidson[416] state that U.S. Department of Health, Education, and Welfare data for 535 analyses of waters from the major watersheds of the United States over a 4-year period showed only two samples containing more than a detectable (10 µg/liter) amount of selenium, the higher of the two being 14 µg/liter. Taylor[746] reports a maximum of 10 samples with a mean of 8 µg/liter in 194 public, finished water-supply sources sampled over a 2-year period. Although Smith and Westfall[714] did not find measurable amounts of the element in drinking waters from 34 of 44 wells in a seleniferous area of South Dakota, the remaining 10 wells did contain from 50 to 330 µg/liter. Hadjimarkos and Bonhorst[293] found averages of 2, 1, and less than 1 µg/liter for 21, 23, and 28 farm samples from three Oregon counties. Samples from 22 Australian villages contained less than 1 µg/liter,[176] and tap and mineral waters from Stuttgart, Germany, have been reported to contain 1.6 and 5.3 µg/liter, respectively.[556] Others have reported some higher values for river waters where irrigation drainage from seleniferous soils has contained up to 2,680 µg/liter of the element.[96,826] For instance, tributaries of the Colorado River receiving such drainage contained up to 400 µg/liter, and the Colorado River itself, below where this water entered it, contained up to 30 µg/liter.

Irrigation Waters, Springs, Wells, Sewage

In addition to high selenium content in irrigation drainage waters, seeps,[36] springs,[36,96,488,498] and shallow wells[93,94,151,498] have been found to contain over 100 µg/liter of the element, but waters in deep wells seem to contain only a few micrograms per liter.[93,94] Beath[35] stated that preliminary tests on Wyoming well waters showed a few instances where enough selenium was present to be poisonous to man or livestock. He also reported that selenate in well water on a Ute Indian reservation apparently caused loss of hair and nails in children,[674] but the evidence was not convincing. Sewage plant effluents contribute to the selenium content of water, as much as 280 µg/liter having been reported in raw sewage, 45 µg/liter in primary effluent, and 50 µg/liter in secondary effluent.[32]

Oceans

Lakin and Davidson[416] summarized the data of Schutz and Turekian, who estimated an average value of 0.09 µg/liter for our ocean waters. Others have found values of 6 µg/liter or less for ocean waters from a

Occurrence

number of locations.[96,263,364,365,415,732] These low levels have been explained by the precipitation of selenite with oxides of metals, such as iron and manganese.[96,263] The mechanism of this precipitation has been studied by a number of investigators.[349,564,568,732,827] Under some conditions, selenite seems to be completely adsorbed in rather high amounts by ferric (and to a lesser extent by aluminum) hydroxide, while selenate is not. The adsorption of the selenite cannot, therefore, be entirely described by the well-known adsorption equation $x/m = kc^n$.

Lakes

Waters in small lakes, in undrained basins into which drainage from seleniferous soils flows,[40] or in stock dams in seleniferous areas[1] have also been found to contain surprisingly little selenium, and again precipitation with metal hydroxides may account for this. However, microbial reduction and precipitation as the element[439] or other biologic mechanisms[1] must also be considered. Selenium has been found in a variety of deep sea deposits,[175,263,415,509,737,828,830] and this too may indicate removal of the element by precipitation of some type.

Precipitation removes selenium from the atmosphere, but reports on quantitative measurements in rain and snow are very limited. Those that have been reported fall between 0.04 and 1.40 µg/liter of water.[323]

Importance of Waters

Apparently, waters rarely contain selenium at levels above a few micrograms per liter. Hence, they can rarely be considered a significant source of the element from either a nutritional or a toxicity standpoint. However, even at the very low levels found in rivers, the large volumes of water involved mean the transport of rather large amounts of the element. Bertine and Goldberg[47] estimated that river flow deposits about 8,000 tons of selenium per year in our oceans. Geologically, therefore, water is important in actively and continuously leaching, transporting, and redepositing the element.

Present U.S. standards for drinking water list 10 µg of selenium per liter as the upper acceptable limit. Pletnikova[611] suggested a limit of 1 µg of selenium per liter as the upper limit for Russian drinking water as a result of his observations on rats. However, the meaning of some of his observations in terms of animal well-being and the diet during treatment is not clear, and, in view of the well-established requirement of animals for selenium, this limit seems unnecessarily low.

AIR

Volcanic Sources, Soil, Plants, Animals

There are a number of sources for selenium in the atmosphere. The element has been found in volcanic gases,[97,737] and if volcanos are, as has been theorized,[96,97] the main source of the element in highly seleniferous sediments, they may be a major contributor of selenium to the air. The occurrence of volatile selenium in plants, particularly in some of the accumulators, has been well documented. Dimethyl selenide[441] and, to a lesser extent, dimethyl diselenide[185] have been identified in volatiles from accumulators. Soils may also contribute selenium to the air as the result of microbial action within them[2] or perhaps because of dusts derived from seleniferous areas. Animals, too, volatilize the element,[661] probably as dimethyl selenide.[474] At present, reasonably accurate estimates of the quantities contributed to the air by each of the above sources are impossible.

Industrial Sources

Dudley[163] summarized potential industrial sources of atmospheric selenium, and these have recently been more thoroughly reviewed.[149] The findings of this review are summarized in Chapter 4.

Concentration in Air

Data on the actual presence of selenium in the atmosphere are limited.[800] In a plant producing selenium rectifiers, air analysis revealed between 0.007 and 0.05 mg/m^3.[15] Selyankina[676] measured concentrations of selenium in the air near two electrolytic copper plants. At one, the concentration was found to be 0.50 μg/m^3; 2 km from the plant, it was 0.07 μg/m^3. At the other, the concentration was found to be 0.39 μg/m^3; 2 km from the plant, none could be detected. Seven air samples collected in the spring at Cambridge, Massachusetts, contained an average of 0.001 μg/m^3 as measured by neutron-activation analysis with chemical separation.[323] Rainwater or snow water collected during a period of 2 years (22 times) contained an average of 0.2 μg/liter. Lakin and Byers[415] reported on the selenium content of some city dusts, finding values between 0.05 and 10 ppm for various cities, but stated that they had no basis for estimating the concentration of the element in the air from their data. Using nondestructive neutron-activation analysis, Dams et al.[141] found values of 0.0025 μg/m^3 at Niles, Michigan, and 0.0038 μg/m^3 at East

Chicago, Indiana, for selenium in suspended particulates in the air. In a related study, values of 0.0008–0.0044 μg/m^3 were reported for particulate matter of the air.[317] Pillay et al.[609] analyzed 18 samples collected around Buffalo, New York, during 1968–1969. These samples consisted of particulates collected on filter paper and gaseous materials absorbed by a liquid trap. They used neutron activation with chemical separation and reported values ranging between 3.7×10^{-3} and 9.7×10^{-3} and averaging 6.1×10^{-3} μg/m^3. Half of the selenium was in the gaseous fraction and half was in the particulate matter collected.

Increasingly, data come from multielement analysis by neutron activation without chemical separation. Care should probably be used in the acceptance of some of the early data for selenium obtained in this way, since there were potential sources of error with the procedures used.[571,634]

The Japanese Association of Industrial Health has recommended a permissible level for selenium compounds in air of 0.1 mg/m^3 (as selenium). This value is a time-weighted average for an 8-hr normal working day and a 40-hr week,[621] and it pertains to confined areas. The USSR standard is also 0.1 mg/m^3; a limit for selenium compounds in workroom air in the United States has been recommended as 0.2 mg/m^3.[15]

Cooper[128] reviewed reports of selenium toxicosis in men working in certain industries where the element is processed. Again, analytical data are sparse, but the situations described suggest that the high selenium levels in these instances could easily be prevented by taking simple precautionary measures.

Rancitelli et al.[618] measured the concentrations of 19 elements in rainwater samples and, to establish the origin of each element, compared these with concentrations in seawater and the earth's crust. They concluded that selenium in the atmosphere does not come from the land or the ocean, but probably results from man's activities, including the burning of fossil fuel. However, volcanic activity and several other sources of atmospheric selenium were not taken into account in arriving at this conclusion. Zoller et al.[844] studied the enrichment values of selenium in atmospheric particles over Antarctica but were unable to tie it to any particular source.

In spite of the paucity of data, it appears that selenium continuously enters and is removed from the atmosphere and that its average concentration in air is very low, probably well below 0.01 μg/m^3. Its chemical form has not been ascertained, but probably a large proportion of what is present is in the particulate matter. The evidence is not sufficient to allow an estimate of what proportion derives from industrial or other man-made sources. Nevertheless, it seems unlikely that pollution of the atmosphere by selenium at present constitutes a problem.

3

Industrial and Agricultural Uses

INDUSTRIAL

Data and text relating to supply and demand of selenium were derived from various U.S. Bureau of Mines published[3,770-780] and unpublished reports.

Production Methods

Although selenium is distributed widely in nature, the tenor of known deposits is insufficient to permit their being mined for selenium alone. Nearly all primary selenium is produced from copper refinery slimes, and current production technology consists mainly of methods for extracting selenium from these slimes. The processes used are primarily designed for effective recovery of precious metals, and selenium recoveries have secondary importance, which is reflected in the low recovery achieved for it.

The first step in slime processing is decopperization. The copper content of slimes ranges from 10% to 70% and is in a form insoluble in cold sulfuric acid. Some slimes can be decopperized with sulfuric acid and steam, but usually roasting is needed to oxidize the contained copper to soluble compounds that can be leached. After the leached residues are smelted in doré furnaces, the metal continues to further refining steps, and the

Industrial and Agricultural Uses

slags are returned to the anode furnace after extraction of by-products.

Selenium may be recovered by volatilization during roasting, by leaching of roasted calcine, by volatilization during furnacing, and by leaching of furnace slag. Slimes containing moderate quantities of selenium and low copper may, following decopperization, be roasted with sodium carbonate flux to form a calcine containing soluble sodium selenite, which is leached with water. Slimes containing moderate quantities of selenium and copper may be roasted with a flux of sulfuric acid and sodium sulfate. Selenium is volatilized as an oxide and scrubbed from the roaster exhaust gas. Slimes having relatively low selenium and copper content may be treated by decopperization and doré furnace smelting. Two methods are used. In one, sodium carbonate flux is added to the slime, and the slag formed contains sodium selenite, which is recovered by leaching. In the other method, the slimes are first smelted with appropriate fluxes. Most of the selenium remains in the metal portion of the melt, which is now refluxed with sodium carbonate to form a slag rich in sodium selenite, which is recovered by leaching. In both methods much of the selenium is volatilized during furnacing and is recovered from the flue gases. All processes use sulfur dioxide to precipitate selenium metal from solutions of sodium selenite and selenious acid.

High-purity selenium is made by several methods, including fractional condensation of volatilized selenium, zone refining, reduction and precipitation from purified selenious acid, and gaseous or wet reduction of purified selenium dioxide.

Some relationships between selenium and refined copper production are shown for Canada and the United States in Table 3-1. The much larger recovery of selenium per unit of copper for Canada is due to the relatively high selenium content of ores from the mines at Noranda, Quebec; Flin Flon, Manitoba; and Sudbury, Ontario. Annual variability of the ratios in both countries was probably a combination of changes in the ore tenor, delays in processing of the residues, and economics of recovery.

Production Levels

The free world refinery production of selenium from 1964 through 1973, as shown in Table 3-2, averaged 2.3 million lb annually. Output during this period has trended upward and has ranged from a low of 1.7 million lb in 1965 to a high of 2.9 million lb in 1970. The United States has been the leading producer for most of these years, followed by Canada, Japan, and Sweden.

TABLE 3-1 Primary Selenium and Copper Production Relationships for Canada and the United States[a]

	Canadian Production			U.S. Production		
Year	Selenium (1,000 lb)	Refined Copper (1,000 tons)	Se/Cu Ratio (lb/ton)	Selenium (1,000 lb)	Refined Copper (1,000 tons)	Se/Cu Ratio (lb/ton)
1964	466	408	1.14	929	1,656	0.56
1965	512	434	1.18	540	1,712	0.32
1966	575	433	1.33	620	1,711	0.36
1967	752	500	1.50	598	1,133	0.53
1968	636	524	1.21	633	1,437	0.44
1969	599	449	1.33	1,247	1,743	0.72
1970	663	543	1.22	1,005	1,765	0.57
1971	718	526	1.37	657	1,592	0.41
1972	582	547	1.06	769	1,873	0.41
1973	598	549	1.09	627	1,868	0.34
TOTAL	6,101	4,913	1.24	7,625	16,490	0.46

[a] Compiled from Bureau of Mines data.[770-780]

In the United States the 1973 production of selenium was accounted for by four concerns with selenium refineries at four copper refineries, as follows:

Company	Plant Location
AMAX, Inc. (formerly American Metals Climax)	Carteret, New Jersey
ASARCO, Inc. (formerly American Smelting and Refining Company)	Baltimore, Maryland
International Smelting and Refining Company (Anaconda)	Perth Amboy, New Jersey
Kennecott Copper Corporation	Magna, Utah

Selenium is recovered in these four plants from slimes generated at their refineries and from interplant transfers of selenium-bearing materials from other domestic and foreign plants.

AMAX, Inc., produces only commercial-grade selenium. ASARCO, Inc., produces commercial and high-purity grades and ferroalloys. International Smelting and Refining Company produces only commercial grade. Kennecott Copper Corporation produces a plus 93%, a commercial, and a high-purity grade.

TABLE 3-2 Selenium: Free World Refinery Production, by Country (1,000 lb)[a]

Country[b]	1964	1965	1966	1967	1968	1969	1970	1971	1972	1973
Australia[c]	4	5	4	4	4	4	7	7	7	8
Belgium–Luxembourg[d]	87	93	91	90	54	46	68	120	147	106
Canada	466	512	575	752	636	820	854	886	655	598
Finland	15	13	12	15	16	14	15	14	16	12
Japan	326	348	421	422	399	435	467	524	738	789
Mexico[e]	7	18	4	—	2	65	278	115	97	86
Peru	17	19	13	11	13	15	15	16	18	18
Sweden	181	176	154	158	168	168	139[c]	134	140	120
United States	899	510	590	568	603	1217	975	627	739	627
Yugoslavia	8	17	21	10	21	20	35	54	55	94
TOTAL	2,010	1,711	1,885	2,030	1,916	2,804	2,853	2,497	2,612	2,458

[a] Compiled from Bureau of Mines data.[770-780] Insofar as possible, data relate to refinery output of elemental selenium only; thus, countries that produce selenium in copper ores and concentrates, blister copper, and/or refinery residues, but do not recover elemental selenium, have been excluded to avoid double counting.
[b] In addition to the countries listed, West Germany and the USSR are known to produce refined selenium, and Zaire and Zambia may produce refined selenium, but available information is inadequate for making reliable estimates of output.
[c] Estimate.
[d] Exports.
[e] Elemental selenium only; excludes selenium content of sodium selenate produced (371,000 lb in 1969).

TABLE 3-3 Selenium Supply–Demand Relationships, 1964–1973 (1,000 lb)[a]

	1964	1965	1966	1967	1968	1969	1970	1971	1972	1973
World production										
United States	899	510	590	568	603	1,217	975	627	739	627
Rest of world	1,673	1,759	1,853	2,040	1,936	2,046	1,892	2,489	2,393	2,376
Total	2,572	2,269	2,443	2,608	2,539	3,263	2,867	3,146	3,132	3,003
Components of U.S. supply										
Refinery production										
Primary	899	510	590	568	603	1,217	975	627	739	627
Secondary	30	30	30	30	30	30	30	30	30	30
Government releases	—	—	—	—	—	—	—	—	14	229
Imports of refined selenium	293	251	286	301	583	546	454	395	430	553
Industry stocks, Jan. 1	1,022	1,305	1,021	797	736[b]	428	240	189	182	161
Total U.S. supply	2,244	2,096	1,927	1,696	1,952	2,221	1,699	1,241	1,395	1,600
Distribution										
Industry stocks, Dec. 31	1,305	1,021	797	445	428	240	189	182	161	106
Exports	100	100	100	196	405	500	376	150	220	264
Government purchases	18	18	—	22	49	—	—	—	—	—
Industrial demand	821	957	1,030	1,033	1,070	1,481	1,134	909	1,014	1,230
U.S. demand pattern										
Electronic components	285	335	385	438	500	555	500	394	458	554
Ceramics and glass	250	349	300	321	300	550	370	316	344	418
Chemicals	186	173	175	169	150	200	135	128	136	160
Other	100	100	170	105	120	176	129	71	76	98
U.S. primary demand (industrial demand less secondary)	791	927	1,000	1,003	1,040	1,451	1,104	879	984	1,200

[a] Compiled from Bureau of Mines data.
[b] Includes a stock adjustment of plus 291.

Industrial and Agricultural Uses

Supply and Demand

Table 3-3 shows the U.S. selenium supply-demand relationships for the period 1964-1973. The elements of production, imports, industry stocks, and government purchases are reported quantities. These components, plus estimated exports, are used to calculate an apparent industrial demand. The distribution of this demand into a use pattern was based on a judgmental balance of diverse sources of information.

The apparent annual consumption of selenium in the United States increased 50% from 1964 to 1973. The most significant increase has been in its use in electronic components, which has risen from 285,000 lb in 1964 to 540,000 lb in 1973, equal to 45% of 1973 demand. The use of selenium in manufacturing glass and allied products, probably its oldest application, increased 67% over the decade because of the increasing quantities of selenium-containing tinted glass used in the construction and transportation industries.

Canada, the source of most of the refined selenium imported into the United States, supplied 516,000 lb, or 93%, of the total imports in 1973.

Uses

Table 3-4 lists the uses of the principal commercial selenium compounds. Some additional compounds that may have commercial uses are ammonium hydroselenate, ammonium selenate, antimony triselenide, arsenic pentaselenide, arsenic triselenide, beryllium selenate, cadmium selenate, cesium selenate, chloroselenic acid, cupric hydroselenite, cupric selenite, gold selenate, gold selenide, hydrogen selenide, lead selenate, lead selenite, lithium selenate, lithium selenite, manganese selenate, manganese selenide, manganese selenite, phosphorus pentaselenide, phosphorus triselenide, potassium biselenite, potassium selenide, rubidium selenate, selenic acid, selenium chloride, selenium oxychloride, selenium oxyfluoride, selenium tetrabromide, selenium tetrachloride, silver selenide, silver selenite, sodium hydroselenite, sodium selenide, stannic selenite, stannous selenide, strontium selenate, strontium selenide, thallium selenate, thallium selenide, and zinc selenate. The ensuing description of industrial uses largely summarizes a more detailed coverage of the subject contained in three publications.[422,731,843] These publications also describe the physical, electrical, and chemical properties underlying the utilization of selenium.

TABLE 3-4 Uses of Some Inorganic Selenium Compounds

Aluminum selenide, Al_2Se_3	In preparation of hydrogen selenide; in semiconductor research
Ammonium selenite, $(NH_4)_2SeO_3$	In manufacture of red glass; as reagent for alkaloids
Arsenic hemiselenide, As_2Se	In manufacture of glass
Bismuth selenide, Bi_2Se_3	In semiconductor research
Cadmium selenide, CdSe	In photoconductors, semiconductors, photoelectric cells, and rectifiers; in phosphors
Calcium selenide, CaSe	In electron emitters
Cupric selenate, $CuSeO_4$	In coloring Cu or Cu alloys black
Cupric selenide, CuSe	As catalyst in Kjeldahl digestions; in semiconductors
Indium selenide, InSe	In semiconductor research
Potassium selenate, K_2SeO_4	As reagent
Selenium disulfide, SeS_2	In remedies for eczemas and fungus infections in dogs and cats; as antidandruff agent in shampoos for human use; usually employed as a mixture with the monosulfide
Selenium hexafluoride, SeF_6	As gaseous electric insulator
Selenium monosulfide, SeS	Topically against eczemas, fungus infections, demodectic mange, fleabites in small animals; usually employed as a mixture with the disulfide
Selenium dioxide, SeO_2	In the manufacture of other selenium compounds; as a reagent for alkaloids
Sodium selenate, Na_2SeO_4	As veterinary therapeutic agent
Sodium selenite, Na_2SeO_3	In removing green color from glass during its manufacture; as veterinary therapeutic agent

ELECTRONIC

Electronic applications account for a substantial part of selenium consumption. Selenium has been used in dry-plate rectifiers, which change alternating current to direct current, for many years. The plates range in size from 8 in. across to a miniature encapsulated type smaller than a match head. About 1952, silicon and germanium rectifiers were introduced. They have since captured a large share of the rectifier market from selenium.

A large and growing electronic use of selenium is xerography, a dry photographic process, which employs metal drums coated with photoconducting amorphous selenium, from which the photographic image is transferred by static electricity. The photoconducting property of amorphous selenium is also the basis of the vidicon television camera. The illuminated portions of the pattern projected upon the selenium layer transmit a light signal when scanned by the electronic beam.

Industrial and Agricultural Uses

Selenium is used in construction of the photoelectric cell. In commercial application, the cell consists of an emitter (a metal surface covered with a thin layer of selenium) and a collector, both enclosed in an evacuated container. It requires an external source of electromotive force, and its output cannot be readily amplified. It has the disadvantage of variability, which makes it unsuitable for precision instruments such as colorimeters and pyrometers. The principal application has been for construction of the electric eye.

The photocell, or photovoltaic cell, utilizes the property of selenium to convert light energy directly into electrical energy. The functioning of devices such as the photographic exposure meter depends on this phenomenon. Solar batteries may be considered photovoltaic cells designed for maximum conversion of solar radiations into electrical energy.

METALLURGIC

Selenium has been used as a degasifier in stainless steel since the mid-thirties. Use for that purpose disclosed that the selenium also improved casting, forging, and machinability properties without reducing corrosion resistance or malleability. The selenium content of casting steel alloys ranges from 0.01% to 0.05%, forging steels from 0.18% to 0.22%, and free-machining steels from 0.05% to 0.35%. Selenium is also added to copper to improve machinability properties.

Since 1959, selenium has been used in chromium plating solutions to produce a plating with better characteristics, to reduce cost, and to improve quality control. Chromium, plated from solutions containing selenate ions, is characterized by about 1,500 microcracks per linear inch. The most desirable pattern is obtained by using a plating solution with 0.012–0.020 g of selenate per liter. Microcracked chromium reduces corrosion of the substrate and provides a surface with less glare.

GLASS AND CERAMICS

The glass and ceramics industry is one of the oldest and largest users of selenium. Selenium is added to the glass melt as elemental selenium, sodium selenate, barium selenite, or sodium selenite in quantities from 0.02 to 0.3 lb/ton to neutralize the green tint in glass due to iron impurities and thus permit manufacture of clear glass. A desirable pink tinge is given to glass by using more selenium. Addition of larger quantities of selenium yields a ruby red glass used in tableware, light filters, and traffic and signal lenses. A large and growing use is in dark-colored glass placed in buildings and vehicles to reduce glare and the rate of heat transfer into air-conditioned spaces.

Mixtures of selenium and arsenic are used in making low-melting glasses having infrared transmitting characteristics.

PIGMENTS

The chemical industry uses an estimated one-eighth of the selenium consumed. Much of this is consumed in pigment manufacture. In the preparation of selenium-containing pigments, the selenium is compounded with cadmium sulfide to obtain the orange–red–maroon cadmium sulfoselenide pigments. These pigments have considerable stability when exposed to sunlight, heat, and chemical attack. They are used to color plastics, paints, enamels, inks, and rubber.

PHARMACEUTICALS

The chemical industry uses selenium as a catalyst in the preparation of pharmaceuticals, such as niacin and cortisone. Selenium is a constituent of fungicides for the control of dandruff and dermatitis. A commercial dandruff shampoo containing selenium sulfide has become an important end use for selenium in recent years.

MISCELLANEOUS USES

Selenium, along with tellurium, is used as an oxidant in the delay train of millisecond-delay electric blasting caps. The quantity of oxidant is dependent on the delay desired but averages about half a gram a cap.

Selenium is used as an accelerator and vulcanizing agent in rubber products to promote heat, oxidation, and abrasion resistance and also to increase the resilience of rubber. Selenium dioxide is used to oxidize, hydrogenate, or dehydrogenate organic compounds. Selenium catalyst hardens fats for use in soaps, waxes, edible fats, and plastics. Selenium also imparts exceptional antioxidant properties to printing ink and mineral, transformer, and vegetable oils and nondrying properties to linseed, oiticica, and tung oils. Selenium oxychloride is a powerful solvent that may be used as a paint and varnish remover and as a solvent for rubber resins, glue, and other organic substances. Selenium compounds find application in lubricating oils and in extreme-pressure lubricants through their antioxidant and antigalling properties; in photographic photosensitizers and toners; and in mercury vapor detectors, fireproofing agents, insect repellents, phosphorescents, and luminescents.

Industrial and Agricultural Uses

AGRICULTURAL

Pesticides

In agriculture, selenium was first used as an ingredient in various compounds for control of mites and insects. Smith[710] reviewed this aspect for the period up to 1961.

About 1933, selenium was used in a material called Selocide, which was prepared by dissolving elemental selenium in potassium ammonium sulfide solutions in proportions corresponding to the formula $[K(NH_4)S]_5Se$. This apparently was the first use of selenium as a pesticide. Selocide was found to be quite effective as a miticide and was used on citrus, grapes, and ornamentals. Later, resistance to Selocide appeared in some types of mites, and the use of this material declined.

Selenium was also used in foliar sprays and in soil applications of selenates. Applications of selenate to the soil resulted in plants that contained concentrations of selenium high enough to render them toxic to certain insects.

Concern over the possible health hazards from selenium residues in plants resulted in restrictions on selenium insecticides. They were restricted first to ornamental plants, then to greenhouse-produced ornamentals.

At the time of Smith's review in 1961, the only registered uses of selenium insecticides were for ornamentals and for use of Selocide on citrus in California. Although there are no records of injury to human beings and animals from the use of selenium-bearing insecticides, it appears that these materials are no longer used.

Control of Dermatitis, Pruritis, Mange

An additional use of selenium, probably stemming from the use of a material called Selsun for control of dandruff in human beings, consists of topical application of a 1% solution of selenium sulfide for control of dermatitis, pruritis, and mange in dogs.

Supplementation for Selenium-Deficient Areas

Discovery of the beneficial effects of selenium in the prevention of certain economically important diseases of livestock led to interest in methods of supplementing livestock with this element. Alternative methods of controlling selenium supplies to livestock are reviewed in *Selenium in Nutrition*.[535]

INJECTABLE METHODS

The use of injectable selenium, usually as sterile mixtures of sodium selenite and vitamin E in an oil base, was the first method of supplementing livestock with selenium to gain general acceptance. This was based on the work of Muth[523] and of Kuttler and Marble.[412] Experiments on various methods and rates of selenium administration have been described by Hartley.[319]

Injectable selenium is normally used on lambs and calves, levels being adjusted to about 1 mg of selenium per 100 lb liveweight. Animals are injected as soon after birth as possible and may be reinjected 4-6 weeks later if it is necessary to maintain their dams on diets low in vitamin E for long periods. Many stockmen in the low-selenium regions of the United States routinely inject all lambs and calves born during the winter months, but they do not inject animals after they and their dams obtain access to green pasture, because green pasture is considered to contain sufficient vitamin E to lower selenium requirements to the point of avoiding death losses. In New Zealand, injections or oral drenches of selenium are used even though animals have access to green pastures. In the United States, injectable selenium formulations are licensed for sheep, lambs, calves, cattle, horses, and dogs, and a selenium capsule is licensed for oral application to dogs. The exact extent of use of injectable selenium is not known. However, it is unquestionably very common in the selenium-deficient areas of the northwestern and northeastern parts of the United States and in adjacent areas of Canada. One authority has estimated that the discovery of the responses of livestock to selenium "will add ten million dollars to the income of livestock producers in Northwest U.S.A. alone."[522]

ADDITIVE TO ANIMAL FEEDS

Decisions concerning the use of selenium as an additive to animal feeds were delayed for reconsideration of the question of carcinogenicity of selenium. On April 17, 1973, in response to a petition of the American Feed Manufacturers Association, the Commissioner of Food and Drugs, U.S. Department of Health, Education, and Welfare, proposed that the food additive regulations be amended to provide for the safe use of selenium as a nutrient in the feed of chickens, turkeys, and swine.[783] The proposed regulation was approved.[782] It provides for the addition of sodium selenite or sodium selenate up to 0.1 ppm of selenium in the complete diet for growing chickens and swine and 0.2 ppm in the complete diet for turkeys. The selenium must be added in a premix formulated in such a way that at least 1 lb but not more than 2 lb of premix are added per ton

Industrial and Agricultural Uses

of complete feed. Feeds containing added selenium may not be administered to laying hens.

In announcing the proposed change in regulations, the Food and Drug Administration provided information relative to the need for selenium supplementation and the safety of this practice. It also stated that "available data [on the carcinogenicity of selenium] have been evaluated by the Food and Drug Administration and the National Cancer Institute," and continued, "Based on these evaluations, it has been concluded that the judicious administration of Se derivatives to domestic animals would not constitute a carcinogenic risk." The Bureau of Veterinary Medicine of the Food and Drug Administration also submitted an environmental impact statement supporting the safety of the proposed regulations.[781]

APPLICATION TO SOIL AND USE OF FOLIAR SPRAYS

Difficulties involved in the addition of selenium to feeds for cattle and sheep and the labor involved in the use of injections of selenium have prompted consideration of the use of soil applications of selenium as a way of meeting dietary requirements of cattle and sheep for this element. Soil applications would require more selenium than the use of injections or feed additions; so would the application of foliar sprays to forage crops. Supplies of selenium and its compounds are not adequate for any widespread use of this element as a soil application. A number of studies concerning selenium application to soils as a means of increasing the concentration of the element in plants have been reported.[52,53,108,110,146,147,252-254,805,806] It has been shown under experimental conditions that the addition of 2-4 lb of selenium as selenite per acre, incorporated into acid or neutral, medium-textured soils, will provide crops with protective but nontoxic concentrations of selenium for several years. However, inadvertent addition of selenium as selenate at this same rate would result in crops containing acutely toxic concentrations of this element. Application of elemental selenium at the same rate probably would not increase the selenium level in crops to the concentrations required to protect animals from selenium-deficiency diseases. Therefore, soil applications of selenium, as a farm practice, at this time is precluded by the lack of necessary supplies of selenium and by the need for close supervision of the form of selenium used and the rate of application. The use of foliar sprays on growing forage and grain crops is less of a problem in terms of selenium required, but additional research is needed before foliar application of selenium can be recommended as a farm practice.

CHIEF AGRICULTURAL USE OF SELENIUM

Without doubt, the chief agricultural use of selenium, in terms of amounts of selenium involved, has been the feeding of forage crops and feedstuffs that naturally contain protective but nontoxic levels of this element to farm livestock. Although this has been done without any realization on the part of stockmen that selenium nutrition of their animals was involved, it has probably been a major factor in the efficiency of livestock production in the United States. According to Kubota *et al.*,[409] there is a large region in the West Central States where crops normally contain adequate but nontoxic levels of selenium. This region includes some of the major feed-grain-, soybean-, forage-, and livestock-producing areas of the United States. Feed grains, soy protein, and dehydrated alfalfa from this region are shipped to many other parts of the United States and to foreign countries for feeding to animals.

4

Cycling

NATURAL

A number of authors have prepared diagrams and discussed the cycling of selenium in nature. In 1939, Moxon et al.[509] developed a scheme depicting the movement of selenium to illustrate what they felt was the role of the so-called converter plants in nature's cycling of the element. Later, Shrift[689] proposed a biologic pathway based on oxidation and reduction processes. Lakin and Davidson[416] illustrated the geochemical movement of the element, and Allaway et al.[11] discussed its cycling at low levels, illustrating not only the directional but also the quantitative aspects of the movement of selenium in the soil–plant–animal system. More recently, Allaway[8] offered a scheme suggesting the chemical changes that occur in the element as it passes through this system. The movement of high levels of selenium in nature has been summarized diagrammatically by Olson.[563]

In retrospect, the proposed pathways of Moxon and co-workers, of Lakin and Davidson, and of Olson depict macrocycles; the proposed pathway of Shrift and the metabolic pathways discussed below depict microcycles; and the changes suggested by Allaway and co-workers depict a mixture of the two. This section will deal only with the macrocycling of the element.

Macrocycles

Figure 4-1 suggests many of the pathways that appear to be of some significance in the movement of selenium between the air, land, and seas.

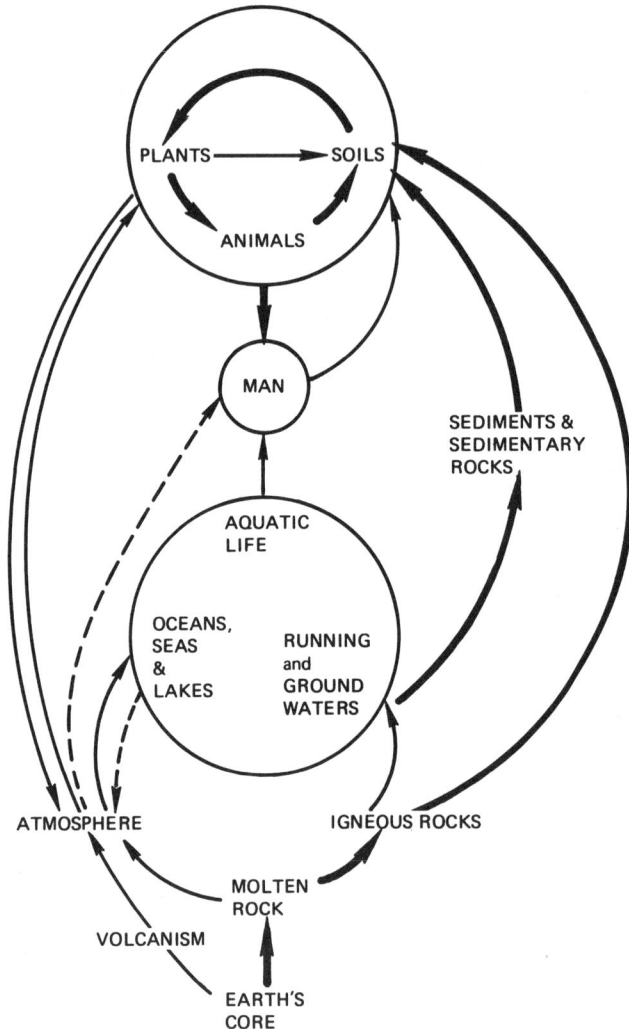

FIGURE 4-1 The cycling of selenium in nature. For simplicity, microorganisms are not included in the above scheme, although they are important to many of the processes involved in the cycle.

While an attempt has been made to quantitate the various paths and to express this quantitation by varying line densities in the figure, data to justify this are very meager or lacking. The evidence that does exist for this scheme is presented in Chapter 2 and in the references cited above. Where direct evidence for pathways is lacking, they are inferred from

Cycling

what we know about the occurrence of the element, the food chain, and geologic and geochemical processes.

EARTH

Selenium seems ubiquitous, and, since the cooling and hardening of the earth's crust, more has been brought to the surface by igneous extrusion and in volcanic emissions. Molten rocks release the element through volatilization into the atmosphere during their cooling, and weathering removes more of the element from the igneous rocks thus formed. Some of the element remains with residual material that eventually forms soils.

ATMOSPHERE

The atmosphere receives selenium from a number of sources other than molten rocks or volcanos. Spray from large bodies of water no doubt contributes some, although the amount is probably very small. Dusts from land surfaces contribute more. Animals exhale volatile selenium compounds, and certain plants produce them and release them into the atmosphere. Microorganisms are in many cases capable of volatilizing the element from a variety of sources. Removal of selenium from the atmosphere in precipitating particulate matter or in rain and snow results in its deposition on land and water surfaces.

WATER

Besides receiving selenium from the atmosphere, oceans and other large bodies of water receive the element from water that runs into them, and they deposit it in their sediments, possibly with the aid of microbial action. Running water also transports the element from drainage areas to floodplains or to poorly drained basins, where it is deposited. Sedimentary rocks or other sediments have been the parent materials for many of our soils.

Selenium and the Food Chain

The cycling of selenium between soils, groundwater, running waters, plants, and animals follows in a general way the cycling of many mineral elements important to the nutrition of man or animals. It appears that soils gradually lose their selenium to eventual deposition in sediments, although some is returned via the atmosphere. It is not possible, actually,

TABLE 4-1 Estimates for Selenium Emission Factors[a]

Mining and milling	
Copper	0.015 lb/thousand tons ore mined
Lead	0.047 lb/thousand tons ore mined
Zinc	0.032 lb/thousand tons ore mined
Phosphate (western)	0.350 lb/thousand tons ore mined
Uranium	0.350 lb/thousand tons ore mined
Smelting and refining	
Copper	0.25 lb/ton copper produced
Lead	0.05 lb/ton lead produced
Zinc	0.04 lb/ton zinc produced
Selenium refining	
Primary (from copper by-products)	277 lb/ton selenium recovered
Secondary	100 lb/ton selenium recovered
End product manufacturing	
Glass and ceramics	700 lb/ton selenium consumed
Electronics and electrical	2 lb/ton selenium consumed
Duplicating	2 lb/ton selenium consumed
Pigments	15 lb/ton selenium consumed
Iron and steel alloys	1,000 lb/ton selenium consumed
Other	10 lb/ton selenium consumed
Other emission sources	
Coal	2.90 lb/1,000 tons coal burned
Oil	0.21 lb/1,000 barrels oil burned
Incineration	0.02 lb/1,000 tons of refuse burned

[a] Derived from W. E. Davis and Associates[149] and related processing data.

to weigh the importance of these two phases of the cycling process and to determine whether selenium is becoming more or less available to the food chain and thus to man. However, as Lakin and Davidson[416] point out, volcanic activity may result in "a land surface enriched in selenium as compared to the earth's crust as a whole."

INDUSTRIAL

A 1972 report by W. E. Davis and Associates[149] for the Environmental Protection Agency (EPA) attempted an inventory of industrial atmospheric emissions of selenium. The report noted the virtual nonexistence of published data on selenium emissions and ascertained through contacts with industry that selenium emissions were not a matter of record. However, estimates for selenium emission factors have been made on the basis of information in the Davis report, other published reports,[608,686,738] and related processing data. The estimates are given in Table 4-1.

Cycling

Mining and Milling

The emission factors for mining and milling were estimated from reports on the selenium content of ores and concentrates at over 40 large mines and mills. Atmospheric emissions from mining and concentrating result mainly from windblown, finely ground tailings. Yearly emissions are estimated at 1% of the selenium placed on tailing dumps annually.

Base Metal Smelting and Refining

Smelting and refining emission factors were estimated from reports at two smelters, from the selenium content of smelter feed, from the amount of metal produced, and from the estimated selenium content in slags.

Selenium Refining

The emission factor of 277 lb/ton of selenium produced in primary selenium refining as done at precious metal refineries is an estimate based on reports of experience obtained from two sources.[149]

Emissions from secondary production of selenium were estimated at 100 lb/ton of selenium produced, on the basis of an office study of processing methods.[149]

End Product Manufacturing

No reliable reports of volatilization losses of selenium in glass manufacturing are available. Emissions are high because the temperature of molten glass is considerably above the boiling point of selenium. The emission factor is estimated at 700 lb/ton of selenium consumed, on the basis of an estimate made for selenium emissions from molten steel of 1,000 lb/ton of selenium added. Emissions from electronic and electrical manufacturing were estimated at 2 lb/ton of selenium consumed, on the basis of information furnished by manufacturers.[149] Relatively small quantities of selenium are emitted to the atmosphere during manufacture of duplicating equipment, according to information obtained from industry. The emission occurs principally during the vacuum-plating process used in manufacturing selenium-coated plates and drums. An estimated 2 lb of selenium was emitted per ton of selenium processed. The major compounders of selenium-containing pigments estimated that 15 lb of selenium was emitted per ton of selenium processed. All reported that bag filters were used for emission control. Selenium emissions during iron and steel alloying are estimated at 1,000 lb/ton of selenium metal consumed.

TABLE 4-2 Estimated Selenium Materials Balance for a Selected Year (1970) (1,000 lb)

	Selenium Input	Atmospheric Emissions	Solid Waste	Selenium Intermediate Product	Selenium in Commercial Product
Production					
Mining and milling	6,400	10	3,600	2,800	0
Smelting and refining	2,800	500	400	1,900	0
Selenium primary refining	1,900	130	800	0	970
Selenium secondary refining	33	3	a	0	30
TOTAL PRODUCTION	—	—	—	—	1,000
Industrial consumption					
Glass and ceramics	370	130	a	0	240
Electronics and duplicating	500	1	a	0	499
Pigments	130	1	a	0	129
Iron and steel alloys	50	25	a	0	25
Other	84	a	a	0	84
TOTAL CONSUMPTION	1,134	—	—	—	—
Other sources and final consumption					
Coal	2,900	1,500	1,400	0	0
Fuel oil	130	130	0	0	0
Incineration	80	a	80	0	0
Other disposal	710	a	700	30	0
TOTAL EMISSIONS	—	2,430	6,980	—	—

a Less than 1,000 lb of selenium.

Cycling

Selenium emissions in other manufacturing processes were assumed to average 10 lb/ton of selenium processed.

Burning Coal and Oil

Estimates of selenium emissions from coal burning are based on an average selenium content of U.S. coal of 2.76 ppm, as reported by Pillay *et al.*[608] and Swanson,[738] and on the results of a study on the disposition of selenium in the combustion products of coal burned in five large modern power plants.[738] The study concerned with the five large plants showed that about 53% of the selenium contained in the coal was emitted to the atmosphere as volatilized selenium or included with particles of fly ash too fine to be trapped by standard dust collectors. In the case of fuel oil, the selenium emission factor was based on analyses of metal concentrations in oil that were done for EPA in 1971. The average selenium content of 10 samples of foreign and domestic crude oil was 0.4 ppm. The average for 27 samples of imported residual oil was 0.6 ppm.

Solid-Waste Incineration

The emission factor for incinerators was based on limited data from a single facility and obviously may not represent a nationwide average. The data were obtained from a 3-day study of an incinerator processing about 245 tons of municipal solid waste daily. Analyses of stack emissions indicated a range of 34–63 lb of selenium per million tons of refuse for the first day and a range of 9–23 lb the second day; none was detected the third day.[378] Household, commercial, and municipal solid wastes are estimated to be in excess of 250 million tons per year, and about 8% of the municipal waste is incinerated.[55] An arithmetic average of the available data indicates the likelihood of relatively small quantities of selenium emissions from incineration of municipal waste.

Selenium Materials Balance

An estimated materials balance for selenium in a selected year (1970) is shown in Table 4-2. It is based mainly on emission factors (discussed in this chapter), on selenium production and consumption (discussed in Chapter 3), and on nonferrous metal production and fuel consumption (information obtained from the Bureau of Mines *Minerals Yearbook*). For selenium production, the input quantity was derived by applying the indicated recovery factor to the reported production. The solid-waste quantities were calculated by subtracting the atmospheric emissions and

TABLE 4-3 Selenium Concentrations in the Atmosphere[a]

	Pyshma			Kyshtym		
Distance from Factory (km)	Number of Samples	Selenium Concentration ($\mu g/m^3$), Mean ± SE	Confidence Limits	Number of Samples	Selenium Concentration ($\mu g/m^3$), Mean ± SE	Confidence Limits
0	16	0.50 ± 0.16	0.18–0.82	18	0.39 ± 0.04	0.31–0.47
0.5	43	0.15 ± 0.03	0.10–0.20	19	0.36 ± 0.05	0.26–0.46
1	24	0.11 ± 0.04	0.07–0.15	18	0.30 ± 0.07	0.16–0.44
2	30	0.07 ± 0.01	0.05–0.08	16	Not detected	Not detected

[a] Derived from Selyankina.[676]

intermediate or commercial products from the input quantities. In the last category of the table, which shows final consumption of selenium, the total quantity incinerated or otherwise disposed of was estimated by assuming that articles containing selenium have an average life of 10 years. The intermediate product recovered from disposed-of material is scrap metal, largely reclaimed photocopying plates.

Although this tabulation represents a reasonable balance for selenium, it must be recognized that it is based on very sparse direct information and in some instances on tenuous assumptions. Additional selenium emissions could derive from pulp and paper manufacture, sulfuric acid production, and other facilities where sulfur is used or contained in the raw materials that are processed.

Selenium as an Industrial Pollutant

In summary, the total industrial emissions of selenium were estimated for 1970 at 2,430,000 lb, or 1,215 tons. Burning of coal accounted for 62% of the total, followed by losses of selenium in nonferrous mining, smelting, and refining operations, which accounted for 26% of the total. Almost all of the remainder was equally divided among precious-metal refinery operations, where all primary selenium is now a by-product; the loss of volatilized metal in glass manufacturing; and the burning of fuel oil.

No serious problem seems to be presented by the relatively small quantity of estimated atmospheric emissions of selenium from industrial sources, assuming adequate dispersal procedures. No reports were found that gave any data on the concentration of selenium emissions resulting from domestic industrial operations or for the selenium content of air near installations that are known to emit selenium. Analyses of selenium in ambient air have been reported for 21 metropolitan areas.[726] All but two of the cities studied were reported to have concentrations of less than 0.04 $\mu g/m^3$ of air. Slightly higher concentrations were reported from Los Angeles and Denver (in the case of Los Angeles, probably due to the dense population and mountainous topography). There seems little measurable correlation between general selenium atmospheric concentrations and the location of industrial selenium emissions. For example, in California very little nonferrous metal is produced and very little coal or fuel oil is burned, but the selenium concentration in the air is about the national average at Long Beach and San Francisco and slightly higher at Los Angeles. It is average at Phoenix, Arizona, and Las Vegas, Nevada. Natural concentrations of selenium appear to be much more important than man-made concentrations, and any industrial selenium pollution would probably be restricted to the immediate vicinity of refineries emitting selenium.

A study has been reported on the atmospheric concentration of selenium in the vicinity of two electrolytic copper refineries in the Ural Mountains of the USSR.[676] Table 4-3 summarizes the results. The selenium was considered to be emitted from the plant treating the anode slimes for by-product recoveries. Effluents from the furnaces were subjected to particulate control considered 96% efficient and then discharged through chimney stacks, 50 m tall at the Pyshma plant and 10 m tall at the Kyshtym plant. Concentrations on the factory grounds were indicated to be largely due to accidental discharges bypassing the stacks.

No data were available to show whether industrial emissions of selenium or selenium compounds to the atmosphere were recycled to the land and sea by normal atmospheric processes. Emissions of sulfur, chemically similar to selenium, are stated to have a life cycle in the atmosphere ranging from 4 hr to 4 days, the time depending on climatic conditions.[21] However, industrial emissions of selenium probably occur primarily as finely divided solid particulates in contrast to sulfur emissions, which are primarily in a gaseous state, implying that selenium emission control requires a different, but possibly simpler, method than does sulfur emission control.[536]

Pollution Control

Reduction of industrial emissions of selenium may occur with implementation of other pollution-control actions. The close association of selenium with sulfur in its occurrence and their similar chemical properties would suggest a corresponding reduction in the emission of selenium as sulfur control and associated removal of particulates become effective. Copper, lead, and zinc smelters were estimated to recover 26% of the sulfur contained in the smelter feed as sulfuric acid in 1970, an increase from the estimated 20% in 1960.[770] Most promulgated emission-control regulations for nonferrous smelters now specify achievement of about 90% sulfur control. Similar sulfur-emission control programs are being instituted for the burning of coal and fuel oil. Improved particulate-control programs at plants processing sludges that contain selenium and at plants consuming selenium would reduce emissions of selenium along with other pollutants.

5

Biologic Effects

METABOLISM

A portrayal of the basic metabolic reactions in which selenium is involved is vital to the understanding of its biologic effects. This section covers metabolic reactions of selenium in animals, plants, and microorganisms.

Animals

The pattern of selenium absorption, distribution, excretion, and retention in animals precedes discussions of mechanisms of bioconversion of selenium and interrelationships with other elements.

ABSORPTION

Gastrointestinal tract The transport of selenium across the intestinal tract was first investigated by Spencer and Blau,[721] who used the everted hamster intestinal sac as a model system. Their studies were limited to selenomethionine, and they found that the selenoanalogue was taken up to about the same extent as methionine. This approach was later expanded by McConnell and Cho,[469] who found that selenomethionine was transported against a concentration gradient, whereas selenite and selenocystine were not. Methionine could inhibit the transport of selenomethionine, but sulfite and cystine did not compete with selenite or selenocystine. The authors suggested in this and in a later paper[468] that a methionine/selenomethionine transport antagonism might account for some of the protective effect of high-protein diets in selenium toxicity.

The gastrointestinal absorption of radioselenite by swine and sheep was studied by Wright and Bell.[835] More selenium was absorbed from the gastrointestinal tract by the monogastric animals than by the ruminant animals. This species difference was thought to be due to the reduction of the administered selenite to insoluble or unavailable forms by rumen microorganisms. Other workers have reached similar conclusions.[88,131,448] Brown et al.[70] recently investigated the effect of dietary selenium on the intestinal absorption of radioselenite by rats. All the animals absorbed from 95% to 100% of the administered ^{75}Se regardless of whether the diets contained 0.02 or 4.02 ppm of selenium. Since the selenium was essentially totally absorbed irrespective of the selenium status of the animal, no absorptive regulatory function appeared to operate under the conditions of their experiment. The authors cautioned, however, that studies on the absorption of the natural forms of selenium that occur in foods are needed.

Lungs Although the poisonous nature of certain volatile selenium compounds, such as hydrogen selenide, is widely appreciated,[165,166] the literature contains practically no quantitative data on the pulmonary absorption of gaseous or finely dispersed particulate selenium compounds. In spite of the fact that elemental selenium is relatively nontoxic, exposure to red selenium fumes causes severe irritation of the mucous membranes when inhaled.[16] A curious difference in the inhalation toxicity of the hexafluorides of sulfur, selenium, and tellurium was noted in rodents.[393] The relative harmlessness of the sulfur derivative was due to the fact that this compound, unlike the selenium and tellurium derivatives, was not hydrolyzed in the lungs to more toxic products.

Skin The literature contains few quantitative data pertaining to the dermal absorption of selenium compounds. However, Dutkiewicz et al.[169] found that 10% of a 0.1 M solution of sodium selenite applied to rat skin was absorbed in 1 hr.

DISTRIBUTION

Internal organs Early studies concerning the distribution of selenium in the internal organs quite naturally dealt with toxic doses of the element, since only the poisonous nature of selenium was appreciated at the time.

Biologic Effects

Also, the methodology for investigating the metabolism of trace quantities of selenium was not available.

Smith et al.[716] found a wide distribution of selenium throughout the body tissues of cats chronically poisoned with subacute doses of sodium selenite. Highest concentrations of selenium were found in the liver, kidneys, pancreas, spleen, heart, and lungs; smaller amounts were found in the red blood cells and brain and still smaller amounts in the plasma, intestine, muscle, bone, and fat. A similar pattern of tissue storage of selenium was seen in animals chronically poisoned with the organic selenium that occurs naturally in grains, although the overall tissue retention of selenium was greater in the animals fed the organic selenium than in those given the inorganic selenite.[715] The distribution of selenium in cattle and sheep chronically poisoned with inorganic selenium was like that reported above for cats.[256,455]

In the experiments referred to above, the animals were exposed to selenium on a long-term basis. Wide distribution of the element in the internal organs of animals was also seen in relatively short-term metabolic experiments conducted by McConnell,[467] who introduced radioselenium as a research tool. After a single subacute dose of radioselenate, the highest concentrations of selenium were found in the liver, kidneys, (total) muscle mass, gastrointestinal tract, and blood. Similar results were reported for the metabolism of radioselenite in mice.[327] The availability of selenium isotopes of high specific activity made possible the study of the fate of microgram quantities of selenium in animals. Hopkins et al.[345] made the surprising observation that the 24-hr percentage distribution of selenium in the internal organs of rats was largely independent of dose over the range 0.025-1 μg of selenium per 100 g of body weight, although carcass retention of selenium was decreased and urinary excretion of selenium was increased at the higher dose level. Likewise, 24 hr after the intraperitoneal injection of 1 μg of ^{75}Se as selenite, there was little or no difference in the amount of ^{75}Se appearing in the liver of rats fed diets containing either 0.04 or 5.04 ppm of selenium. However, a decreased amount of ^{75}Se appeared in the kidneys, blood, and carcass of the animals fed the diet containing the higher level of selenium. This metabolic difference between the liver and the rest of the body was thought to be consistent with a relatively rapid turnover of selenium in the liver. Although only relatively small amounts of ^{75}Se were found in the testes in these short-term studies, long-term experiments have demonstrated the importance of this organ in selenium metabolism.[82]

Since the chemical form of selenium in foods is apt to be organic selenium rather than selenite or selenate, the distribution of organic forms of ^{75}Se is of interest. Jacobsson[367] showed that the uptake of ^{75}Se in the tissues of sheep after injection of a single dose of selenite, selenomethionine, or selenocystine was about the same, except that a higher ^{75}Se level was found in the pancreas when the seleno-amino acids were given. Likewise, Awwad et al.[28] found the most radioactivity after the injection of ^{75}Se-selenomethionine in the pancreas, intestine, liver, and kidneys. Anghileri and Marques[19] also found a high level of activity in the pancreas 8 hr after intravenous injection of ^{75}Se-selenomethionine or ^{75}Se-selenocystine, but this activity tended to disappear from this organ after 1 week, whereas the activity associated with the liver remained constant. Moreover, there was a considerable increase in the ^{75}Se activity in the testes throughout the experiment.

The distribution of selenium in the animal body under normal physiologic conditions was studied by Jones and Godwin,[382,383] who fed mice alfalfa that had been grown in a solution culture containing $H_2{}^{75}SeO_3$. When expressed in terms of activity per unit weight of organs, the concentrations of ^{75}Se ranged in the following order: kidneys > liver > pancreas >> lungs > heart > spleen > skin > brain > carcass. Except for a reversal of the kidneys and liver and the lower amounts of selenium in the spleen, the relative concentrations of selenium in this experiment, performed with normal levels of selenium, correspond closely with the relative concentrations of selenium found by Smith et al.,[716] who were studying chronic selenosis. In a similar study, Jenkins and Hidiroglou[372] exposed dystrophogenic pasture grass to radioselenite and fed extracts of these plants to preruminating lambs. The distribution of ^{75}Se in the internal organs of lambs fed very low levels of selenium was very much like that seen in cats chronically poisoned with selenite.[716]

Blood Selenium rapidly appears in the blood following a single injection of selenate into rats, and the concentration in the plasma is initially greater than that in the erythrocytes.[467] Gradually, however, the plasma loses and the red blood cells gain selenium, so that after 3 hr the formed elements contain more selenium than the plasma. This is in agreement with earlier work showing that the amount of selenium in the erythrocytes was greater than that in the plasma in cats chronically poisoned with selenite.[716] McConnell et al.[479] performed a time study on the distribution of ^{75}Se in the serum proteins of dogs injected with radioselenium and found that within the first hour the greatest activity was in the albumin fraction. After this time, the albumin-bound ^{75}Se decreased, and the activity associated with α_2 and β_1 globulins increased. This shift of ^{75}Se activity from albumin to globulins was essentially confirmed by Jenkins et al.[373] in chicks receiving high doses of radioselenite. In birds given low doses of ^{75}Se, however, the initial binding by albumin was not observed. Rather, the ^{75}Se was first located in the α_2 and α_3 globulins, which then migrated to the α_2 and globulin fractions after 24 hr. The report of McConnell and Levy[472] that ^{75}Se was present in the serum lipoproteins of dogs or rats 24 hr after injection with $H_2{}^{75}SeO_3$ was contradicted by Roffler et al.,[633] who claimed that the activity associated with lipoprotein was due to contamination by other serum proteins of high specific activity.

Jensen et al.[377] showed that the *in vivo* incorporation of an injected dose of ^{75}Se into red blood cells was greater in chicks that had previously been fed a diet deficient in selenium than that in chicks that had been fed a diet supplemented with the element. Wright and Bell[836] then found that sheep erythrocytes took up ^{75}Se *in vitro* via an oxygen-dependent transport mechanism to an extent that was inversely proportional to the dietary intake of selenium. The authors suggested that this technique might have value as an indicator of the selenium status of animals, and this suggestion was confirmed in respect to sheep by the work of Weswig et al.[816] and Lopez et al.[447] Burk et al.[81] measured the *in vitro* uptake of ^{75}Se by red blood cells from children with untreated kwashiorkor and normal children and found that the erythrocytes from the malnourished children took up almost twice as much ^{75}Se as did those from control children. This result agreed with the lower levels of selenium observed in the blood of children with kwashiorkor.

Intracellular The distribution of ^{75}Se in rat intracellular liver fractions was first studied by McConnell and Roth.[477] These workers found 6.2%, 24.4%, 15.3%, and 50.3% of the total ^{75}Se activity of the homogenate in the nuclear, mitochondrial, microsomal, and soluble fractions, respectively, 24 hr after injection with radioselenite. Brown and Burk[69] reported the following percentage distribution of ^{75}Se in the same subcellular fractions of rat liver 24 hr after injection: 22.7%, 4.7%, 22.6%, and 52.7%. The reason for this discrepancy in the nuclear and mitochondrial fractions is not clear but may be related to the fact that Brown and Burk fed their rats diets that were deficient in selenium and administered much smaller doses of radioselenite to their animals. Wright and Mraz[838] showed that the relative proportion of ^{75}Se appearing in the mitochondrial fraction could be increased by feeding diets high in selenium, although this shift of ^{75}Se to the mitochondria occurred at the expense of the soluble rather than the nuclear fraction. Brown and Burk made the additional observation that the relative distribution of ^{75}Se in the subcellular fractions of rat liver or testes seen 24 hr after injection did not change for periods as long as 10 weeks postinjection. There was, however, a specific pattern of subcellular ^{75}Se distribution in each of the organs; the nuclear and mitochondrial fractions of the testes contained relatively higher amounts of ^{75}Se than did the same fractions in the liver. This organ specificity in the subcellular distribution of ^{75}Se was also noted in chicks by Wright and Mraz, who found that the kidney nuclear fractions contained more, and the microsomal and soluble fractions less, ^{75}Se than the corresponding liver fractions.

EXCRETION

Urine The urinary excretion of selenium is the main pathway in chronic selenosis. Cats poisoned with sodium selenite eliminated 50%–80% of the intake by this means, but only 20% or less in the feces.[715] The urinary excretion of selenium by cats poisoned with organic selenium was only about 40% of the intake, but this was due to a greater retention of the selenium rather than to an increased fecal output. The importance of the urinary pathway was also seen in the work of McConnell,[467] who showed that over 40% of a single subtoxic dose of selenate was excreted in the urine in 24 hr, whereas only 3%–6% appeared in the feces. Much smaller amounts of selenium appeared in both urine and feces after the initial 24-hr collection period.

Burk *et al*.[82] recently completed a thorough analysis of the various factors that can influence the urinary excretion of selenium in rats injected with submicrogram quantities of the element. The amount of ^{75}Se appearing in the urine was found to be directly related to the level of selenium in the diet fed the animals. Only about 6% of an administered dose of ^{75}Se was excreted after 10 days by rats fed a diet containing 0.004 ppm of selenium, whereas 67% of the dose was excreted by rats fed a diet containing 1 ppm of selenium. The urinary output of ^{75}Se was also shown to depend directly on the size of the dose of selenium given; addition of 50 μg of nonradioactive selenium carrier increased the amount of ^{75}Se in the urine eightfold. These workers postulated the existence of a threshold dietary level of selenium above which urinary excretion of selenium is directly related to its dietary level and below which it is not. Under their conditions, this threshold lay between 0.054 and 0.084 ppm of dietary selenium.[83]

Feces In rats, the fecal pathway appears to be a constant and rather minor route of elimination of selenium over a wide variety of conditions. Burk *et al*.,[82] for example, injected rats with a tracer dose of $H_2^{75}SeO_3$ (=5 ng of selenium) and found that about 10% of the dose was excreted in the feces in 10 days, regardless of whether the diet fed contained 0.004 or 1 ppm of selenium. Likewise, adding up to 200 μg of carrier to the tracer dose of ^{75}Se had little influence on the percentage of the dose excreted via the feces. Brown *et al*.[70] also found that addition of selenium as carrier or in the diet would not increase the fecal excretion of ^{75}Se above 10% of

Biologic Effects

the dose when the ^{75}Se was given orally. On the other hand, when rats are chronically poisoned with selenium, 20%–50% of the selenium ingested may be excreted in the feces.[264] A striking effect of route of administration on the fecal excretion of selenium in swine was noted by Wright and Bell,[835] who showed that barrows given selenium by injection excreted only 3% of the dose in the feces, whereas animals given selenium orally excreted 15% of the dose in the feces. This effect of route of administration was even more pronounced in sheep; wethers injected with selenium excreted only 5% of the dose in the feces, whereas animals given selenium by mouth excreted 66% of the dose in the feces. This increased fecal excretion of selenium by ruminants was thought to be due to the reduction of the administered selenite to insoluble or unavailable forms by rumen microorganisms; indeed, Butler and Peterson[88] showed that most of the selenium in sheep dung was present in water-insoluble forms.

Lungs The pulmonary excretion of selenium usually attains importance only in subacute selenium toxicity. This is clearly shown from the systematic work of Olson *et al.*,[572] who measured the exhalation of volatile selenium compounds by rats injected with graded dose levels of selenite. Only negligible amounts of volatile selenium were formed when the animals received 0.4 mg of selenium per kg of body weight or less, but when the level of selenium administered reached 1.9 mg of selenium per kg of body weight, about 30% of the dose could be eliminated as volatile selenium. Ganther *et al.*[241] showed that the production of volatile selenium compounds by rats depends on a number of dietary variables. Selenium volatilization could be increased by increasing the protein and methionine content of the diet, by adding 5 ppm of selenium to the diet, or by feeding certain crude diets.

Bile and pancreatic juice Under ordinary circumstances, the biliary pathway is not a significant route for selenium excretion. McConnell and Martin,[473] for example, found only about 2% of an injected dose of selenate in the bile of dogs after 7 hr. On the other hand, the biliary pathway assumes major importance in the detoxification of selenium by arsenic.[431]

Hansson and Blau[311] reported that selenomethionine could be incorporated intact into the pancreatic juice proteins of cats *in vivo*, but analogous studies with other forms of selenium appear to be lacking.

Saliva Hadjimarkos and Shearer[295] have reported that the amount of selenium in the saliva of normal children ranged from 1.1 to 5.2 ppb, with a mean concentration of 3.1 ppb. There appears to be no systematic study on the effect of selenium exposure or dietary selenium on the salivary excretion of selenium.

Hair The amount of selenium in the hair was first suggested as an index of the extent of the deposition of the element in the tissues in chronic selenium poisoning.[565,814] Later the selenium content of hair was used as an indicator for the frequency of nutritional muscular dystrophy in selenium-deficient beef cattle.[333] Analysis of hair samples has proved to be a convenient, reliable diagnostic aid in assessing various trace-element deficiencies in human beings.[304,398,399] McConnell and Kreamer[471] showed that trace amounts of ^{75}Se injected into dogs were deposited and retained in hair for as long as 316 days. The ^{75}Se was found in the cystine-rich protein keratin and in the cystine fraction isolated from hair. Studies with rats and sheep given radioselenite showed a gradual appearance of ^{75}Se in the hair and wool over a period of several weeks.[69,187]

A preliminary study of selenized wool showed that selenium did not significantly affect fiber length, thickness, or strength.[426] However, the selenium content of the wool did have a significant correlation with the number of distorted fibers in the fleece. Holker and Speakman[342] studied the action of selenium dioxide on wool that had been reduced with thioglycolic acid. Selenium-containing cross linkages were formed in which each selenium atom linked two cysteine side chains to form R—S—Se—S—R structures.

Retention Burk et al.[84] studied the tissue selenium levels during the development of dietary liver necrosis in rats fed torula yeast diets and found great drops in the selenium levels of the kidneys, liver, and blood in animals fed low-selenium diets for 4 weeks. Similar results were reported by Hurt et al.[363] for rats fed amino acid diets deficient in selenium. Buchanan-Smith et al.[80] were able to deplete the tissue selenium concentrations in 4-month-old sheep by feeding them a nonprotein purified diet for 6 months. Selenium concentrations in the tissues of swine and lambs with nutritional muscular dystrophy have been found to be lower than those in the tissues of normal animals,[87,445] and Allaway et al.[13] suggested that a concentration of 0.21 ppm of selenium in the livers of lambs was the critical minimal level for the prevention of white muscle disease. Although it is generally accepted that low levels of tissue selenium are needed for the development of selenium-responsive diseases, some studies indicate that such levels may not be the only factor involved.[319]

Blincoe[60] examined the whole-body turnover of ^{75}Se in rats injected with 0.14 mg of selenium as selenite during a 2-week period and found that the turnover could be described by two first-order processes of widely differing rate constants. The author concluded that the selenium was transferred from a rapidly excreted form to a more slowly excreted form. These results were verified by Ewan et al.,[188] who showed that, after a single subacute injection of radioselenite to rats, the selenium was eliminated rapidly for about 3 days and that, after a period of transition, a slow constant rate of loss persisted for months. The amounts of ^{75}Se retained after the rapid-elimination phase varied with the size of the dose, but the rate of loss thereafter did not. A series of diets fed to rats after the fixed pool had been established showed that the rate of depletion from that pool varied with the selenium intake during the depletion period; that is, the long-term loss of ^{75}Se was unaffected by changes in the diet, except as the diets varied in their selenium content. An increase in either dietary or carrier selenium was also shown to decrease the whole-body retention of ^{75}Se in rats injected with only a tracer dose of radioselenite equivalent to 5 ng of selenium.[82] A later study showed that increases in dietary selenium of as little as 0.06 ppm could cause significant decreases in the whole-body retention of tracer doses of ^{75}Se.[83]

The recent decision to allow the addition of inorganic selenium salts as a nutritional supplement to the feed of swine and poultry was made after research had demonstrated that such use of the element would not contribute to an excessive buildup of selenium residues in the edible tissues.[782] Early work with rabbits fed high levels of selenium had shown that the retention of the element in the tissues of animals fed the naturally occurring organic form of selenium in grains was far greater than its retention when administered as the inorganic sodium selenite.[715] Analogous results were obtained more recently in chicks and poults fed nutritional levels of selenium. Scott and Thompson[672] found that those birds fed diets containing 0.67 ppm of selenium added as the organic form in soybean meal or wheat had higher levels of selenium in the blood and muscle than those fed diets containing 0.67 ppm of selenium as selenite. However, not all naturally occurring forms of selenium are retained in the tissues to a greater extent than inorganic selenium; Miller et al.[487] found that the selenium in fishmeal and fish solubles was retained less efficiently by chicks than selenium as selenite. This result would agree with the finding by Scott and Cantor[670] that the selenium in tuna meal is only about 40% as effective as selenite selenium in preventing exudative diathesis in chicks. Additional work by Scott and Thompson[672] showed that the retention of selenium in various chick tissues did not increase appreciably when the selenium content of the diet (as selenite) was raised from 0.2 to 0.4 or 0.8 ppm. Apparently a sensitive control mechanism operates at nutritionally required levels of selenium that allows for the rapid excretion of any excesses of inorganic selenium above the amounts needed by the body. A "plateauing" of tissue selenium levels in chicks receiving graded amounts of dietary selenite was also observed by Scott and Cantor.[670]

Similar results concerning selenium residues in swine tissues have been obtained. The level of selenium in pork muscle was found to be directly related to the level of dietary selenium when the element was supplied in the naturally occurring organic form,[406] whereas supplementation of a practical swine diet low in selenium with 0.2 ppm of selenium as selenite caused only a modest increase in the level of the element in muscle, which then returned to baseline (deficient) levels 60 days after withdrawal of the supplemented feed.[272] Experiments with steers and sheep given doses of inorganic selenium at levels needed to prevent white muscle disease indicated that the residues of selenium in tissues returned to background levels after depletion periods of 15 weeks (steers) and 21 weeks (sheep).[336,413] The studies outlined above lead to the overall conclusion that, when animals are supplemented with nutritional amounts of inorganic selenium, there is little or no tendency for selenium to accumulate in the edible tissues of the animals above the levels that are known to occur in animals fed diets containing adequate quantities of naturally occurring selenium.

The regulation allowing the use of selenium as a feed additive did not apply to laying hens. Selenomethionine injected into chickens is known to be incorporated as such into egg white proteins.[554] On the other hand, although feeding 8 ppm of selenium as selenite in a diet for laying hens increased the amount of selenium in eggs to three times the normal amount, the selenium in eggs returned to a normal level 8 days after removal of this highly seleniferous diet.[23]

BIOCONVERSION

Methylated derivatives The circumstances under which significant amounts of volatile selenium compounds are produced by animals have been discussed. The main volatile selenium metabolite exhaled by rats following administration of inorganic selenium was characterized by McConnell and Portman[474] as dimethyl selenide, but it is not known whether other minor volatile selenium metabolites exist. This methylation of selenium was considered to be a highly effective detoxification mechanism; dimethyl selenide was shown to be about one five-hundredth as toxic as selenite.[475] The generally presumed innocuousness of dimethyl selenide, however, may have to be reevaluated in light of its remarkable synergistic toxicity with mercury.[589]

Rosenfeld and Beath[639] showed that several bovine tissues were able

Biologic Effects 61

to convert selenite into volatile selenium compounds *in vitro*. Ganther[233] performed a thorough study on the enzymic synthesis of dimethyl selenide from sodium selenite in mouse liver extracts. The probable methyl donor for this process was shown to be S-adenosyl-L-methionine, and the system had a specific requirement for glutathione. A survey of various tissues for their ability to synthesize dimethyl selenide demonstrated that the liver and kidneys had the highest activity of the tissues studied, lungs had intermediate activity, and leg muscle, spleen, and heart had the lowest activity. The occurrence of volatile selenium compounds in many tissues causes problems for analysts wishing to measure selenium in biologic samples, since drying tissues can cause up to a 25% loss of selenium.[326]

In 1969, two groups independently reported that trimethylselenonium ion, $(CH_3)_3Se^+$, was a urinary metabolite of selenium.[92,584] This was the first chemical characterization of an organic selenium compound that was excreted in the urine. Trimethylselenonium ion appeared to be the main excretory product of selenium metabolism, since it routinely accounted for 30%-50% of the urinary selenium, regardless of whether high or low doses of selenium were given. This compound also appeared to be a general excretory product from selenium metabolism, since it was the predominant metabolite, no matter which form of selenium was given to the animal.[585] Trimethylselenonium ion may also represent another example of a methylated detoxification compound of selenium, as suggested by Byard,[92] since it is much less toxic than selenite or selenate.[553] The relative biologic inactivity of trimethylselenonium ion was demonstrated in yet another way when this selenium derivative was shown to be ineffective in preventing dietary liver necrosis.[764] However, trimethylselenonium ion does have a curious synergistic toxicity with arsenic.[553] Although trimethylselenonium ion appears to represent the main urinary metabolite of selenium in a wide variety of conditions, other uncharacterized excretory products of selenium exist.[91,585]

The nature of selenium in tissue proteins   During the 1930's, studies conducted to determine the chemical nature of selenium in the tissues of animals poisoned with selenium showed that a large percentage of the selenium in the internal organs remained in the protein fraction after extraction with trichloroacetic acid and bromine–hydrobromic acid.[715] This was particularly true in the liver; from 30% to 94% of the total selenium was found in the liver protein fractions. The exact form of selenium in the tissue proteins, however, was not established by these experiments.

In 1957, this problem was approached by using paper chromatography to characterize the selenium compounds present in the liver protein hydrolysate of dogs 24 hr after injection with ^{75}SeCl$_4$.[478] After ethanol, ether, and trichloroacetic acid extractions, the liver protein residue was hydrolyzed by refluxing in 6N HCl, a procedure now known to destroy both selenomethionine[687] and selenocystine.[351] In spite of the destruction of these seleno-amino acids, chromatography of the hydrolysate still revealed at least three compounds containing radioactive selenium; the greatest activity was in the cystine–selenocystine area, less in the methionine–selenomethionine area, and the least in the leucine area.

In 1964, Schwarz and Sweeney[669] showed that selenite was bound to certain sulfur compounds *in vitro* to give reaction products that had chromatographic mobilities similar to those of the parent sulfur compound. These findings called into question the use of chromatography as the sole criterion of identification of selenium compounds and also prompted Cummins and Martin[138] to reinvestigate the question of whether selenocystine and selenomethionine were synthesized *in vivo* from sodium selenite in mammals. These workers studied the alkaline dialysis of a liver homogenate prepared from a rabbit that had been fed radioselenite for 5 weeks. Large injected and oral doses of radioselenite were also given to the animal 24 hr before sacrifice. Dialysis of the homogenate under alkaline conditions removed 93% of the radioactivity originally in the homogenate, and 90% of the selenium in the dialysate was recovered as selenite. After dialysis, the protein of the liver homogenate was enzymically hydrolyzed. Ion exchange chromatography of the enzymic hydrolysate failed to demonstrate any radioactivity in the vicinity where the seleno-amino acids were known to appear on the chromatogram. These workers also used ion exchange chromatography to characterize the urinary excretion products of a rabbit that had been injected 24 hr previously with radioselenite. Although two distinct radioactive fractions that suggested the presence of selenocystine and selenomethionine were detected in the urine, similar radioactive fractions could also be obtained by adding radioselenite to a normal urine sample *in vitro* or by adding radioselenite to a chemically defined mixture of sulfur compounds. Furthermore, chemical analysis of the radioactive fractions of the urine sample labeled *in vivo* showed that most of the selenium was present as selenite. Two major conclusions arose from this work: (1) there exists no pathway for the *in vivo* synthesis of selenocystine or for selenomethionine from selenite in the rabbit, and (2) sulfur compounds bind selenite to yield complexes that have chromatographic properties similar to seleno-amino acids.

One would have hoped that the experiments of Cummins and Martin would have settled once and for all the important question of whether monogastric animals are able to biosynthesize seleno-amino acids from

Biologic Effects

inorganic selenium, but a recent communication by Godwin and Fuss[259] raises the question anew. These workers used two-stage column chromatography to characterize the selenium compounds in the enzymic hydrolysate of kidney protein from a rabbit injected intravenously with selenite. A very small percentage of the radioselenite given, about 0.5%, apparently was converted into selenocystine. A number of other amino-acid-like selenium compounds were found in the liver, kidneys, and plasma, but no attempt was made to identify them. Clearly, the whole problem of selenium metabolites in tissues is one that requires much additional effort.

If, then, the biosynthesis of seleno-amino acids represents only a minor or nonexistent pathway of selenium metabolism in nonruminant animals, what happens to selenite *in vivo*? One possibility arises from a reaction first postulated by Painter:[578]

$$4RSH + H_2SeO_3 \rightarrow RSSR + RSSeSR + 3H_2O.$$

Ganther[236] has characterized compounds of the RSSeSR family (selenotrisulfides) after reacting selenious acid nonenzymatically with cysteine, 2-mercaptoethanol, or coenzyme A. The reaction of selenious acid with glutathione appeared to be somewhat more complex, and under physiologic conditions of pH and reactant concentrations, the selenotrisulfide derivative of glutathione (GSSeSG) reacted further to form glutathione selenopersulfide:[234]

$$GSSeSG + GSH \rightarrow GSSeH + GSSG.$$

Such selenopersulfide formation may be important in the biologic function of selenium. Proteins containing thiol groups also apparently can undergo selenotrisulfide formation; reduced pancreatic ribonuclease could be cross-linked with selenium to form an intramolecular —S—se—S linkage in place of a disulfide.[235] Jenkins and Hidiroglou[371] found a good correlation between the cysteine content and the selenium uptake capability of certain proteins, and this was taken as additional evidence for selenium incorporation by selenotrisulfide formation.

Although selenotrisulfide formation provides a plausible rationale for the initial binding of selenite by tissues, it should be pointed out that the nature of the binding changes with time in some unknown fashion.[370,485] For example, reduction with thiols, sulfitolysis, or alkali treatment released over 70% of the radioactivity from serum proteins taken from chicks 4 hr after oral dosing with radioselenite. At 96 hr after radioselenite administration, however, much less ^{75}Se was removed by reduction or sulfitolysis, whereas the alkali treatment remained equally effective. The formation of seleno-amino acids was not considered to be a likely reason for this phenomenon, since the average half-life of the chick serum proteins is too long to account for the relatively rapid change in selenium susceptibility to release. A better explanation was thought to be that the strength of binding of selenium could be influenced by the nature of the amino acid residues in the vicinity of the binding site. Then with time the selenium in the more labile binding sites would be lost, and a higher proportion of the more resistant complexes would remain. Whether this theory is correct can be determined only by additional research.

METABOLIC INTERRELATIONSHIPS

Sulfur Halverson and Monty[300] first showed that dietary sulfate could partly counteract the toxic effects of selenium in rats. A later report by Halverson *et al.*[301] demonstrated that this beneficial effect of sulfate in selenium poisoning was specific for selenate in that little or no protection was observed against selenite or organic forms of selenium. These workers also found that sodium sulfate in the diet increased the urinary excretion of selenium from rats fed selenate but had no significant effect on the fecal excretion of selenium. Ganther and Baumann[238] noted that this increased urinary excretion of selenium due to sulfate was accompanied by a decreased retention of selenium in the internal organs. Although Schubert *et al.*[660] obtained an increase in the incidence of white muscle disease in lambs after treating a field of alfalfa with gypsum, Paulson *et al.*[595] did not see any significant effect of dietary sulfate on the fate of a physiologic dose of selenate administered via rumen puncture to lactating ewes.

Some authors have claimed that dietary methionine is of value in alleviating selenium toxicity,[443] whereas others have not found this to be the case.[400,712] A possible rationale for this discrepancy was offered by the results of Sellers *et al.*,[675] who demonstrated that methionine could protect against selenium toxicity but only when adequate levels of vitamin E were present in the diet. Levander and Morris[432] confirmed that methionine and vitamin E protect against the liver damage caused by excess selenium, and they also showed that several fat-soluble antioxidants could replace the vitamin E in potentiating the beneficial response due to methionine. Water-soluble antioxidants, such as ascorbic acid or methylene blue, could not substitute for the vitamin E. The selenium content of the liver and kidneys was significantly decreased in those groups that were fed protective methionine/antioxidant combinations.

Although various sulfur compounds obviously can act to minimize the toxicity of selenium under many conditions, the exact mechanisms by which these sulfur compounds exert their beneficial effects have not been elucidated in detail.

Biologic Effects

Cadmium and mercury   Selenium has the peculiar property of being able to protect animals against the toxic effects of injected subacute doses of cadmium and mercury,[589] and it has been suggested that one of the biologic functions of selenium could be the protection of the organism against the toxicity of trace amounts of metals that even under "normal" conditions enter the body from the environment.[338] This suggestion gained considerable credence when Ganther et al.[240] showed that dietary selenite could decrease the symptoms of chronic methyl mercury poisoning in rats. These workers also speculated that the selenium present in tuna might lessen the danger to man of mercury in tuna. Not all forms of selenium exert a beneficial effect in mercury poisoning, however. A remarkable synergistic toxicity has been observed between the relatively nontoxic dimethyl selenide and certain mercuric salts.[589] The biochemical basis of all these fascinating relationships between selenium and the metals in Group IIB is still largely unknown.

Arsenic   The unusual ability of arsenic to decrease the toxic effects of selenium was discovered by Moxon,[506] who found that 5 ppm of arsenic as sodium arsenite in the drinking water completely counteracted the liver damage in rats caused by 15 ppm of dietary selenium as seleniferous wheat. Ganther and Baumann[237] found that subacute doses of arsenic decreased the exhalation of volatile selenium compounds but increased the excretion of selenium into the gastrointestinal tract. This effect of arsenic on the pulmonary excretion of selenium was thought to be paradoxical, since selenium volatilization was considered the main detoxification pathway in animals injected with subacute doses of selenium. Levander and Baumann[430] noted that, as the dosage of arsenic was increased, the amount of selenium excreted into the intestine went up, but the level of selenium retained in the liver went down. This observation was explained by an enhanced biliary excretion of selenium caused by the arsenic.[431] The bile is normally a relatively minor route of selenium excretion, but in the presence of arsenic the amount of selenium eliminated via the bile can be enhanced tenfold. The molecular mechanism by which arsenic stimulates the biliary excretion of selenium is not known, but selenium in the bile of rats also given arsenic exhibits dialysis behavior much like that of selenium in serum proteins.[431]

Although arsenic in the drinking water can protect against the toxic effects of excess dietary selenium and although injected doses of arsenic can protect against injected doses of selenium, it should not be assumed that arsenic will always elicit a benign response in selenosis. Arsenic added to the diet, for example, has variable effects against selenium poisoning.[104,162,238,504,796] Moreover, if selenium and arsenic are both added in the drinking water, there can be an additive toxicity of the two elements if the amounts given are high enough.[221]

Finally, Obermeyer et al.[553] showed that the toxicity of trimethylselenonium chloride to rats can be increased 20-fold if arsenic is injected along with the selenium compound. More research is needed to clarify these intriguing metabolic interrelationships between selenium and arsenic.

A preliminary report that arsenate given to ewes fed a selenium-deficient ration could decrease the incidence of myopathy in the lambs (Muth et al.[529]) has not been confirmed.*

Linseed meal A survey of several different protein sources revealed that linseed meal has a unique protective activity against chronic selenium poisoning.[505] The protective factor in the linseed meal was not associated with protein and could be extracted with hot aqueous ethanol.[302] Olson and Halverson[567] reported that, although the animals fed linseed meal were largely protected against the toxic effects of the dietary selenium, their internal organs actually contained higher concentrations of selenium than did those of animals fed diets not supplemented with the linseed meal. This somewhat unexpected result was confirmed by Levander et al.,[436] who also showed that the selenium in the liver homogenates of animals fed the meal appeared to be more tightly bound than the hepatic selenium of animals not fed the meal. These results are a clear illustration of the principle that the chemical form of selenium in the tissues may be more important than the concentration present in determining the selenium status of an animal.

Plants

Rosenfeld and Beath[641] divided plants into three groups on the basis of their propensities for accumulating selenium, and the following is a modification of their classification:

Group 1. Plants in this group are referred to as primary selenium accumulators. They contain high amounts of the element (often over 1,000 ppm). The selenium in some species is largely water soluble and appears in compounds of low molecular weight. These plants are often referred to as "indicators," since they appear to grow only on the more highly seleniferous soils. Included are many species of *Astragalus* and some species of *Machaeranthera, Haplopappus,* and *Stanleya.*

---

*P. H. Weswig and P. D. Whanger, unpublished observations.

Biologic Effects

Group 2. Plants in this group are referred to as secondary selenium absorbers. They rarely contain more than a few hundred parts per million of the element. Most of the selenium is in the selenate form; small amounts are in the organic form. Included are many species of *Aster* and some species of *Atriplex, Castelleja, Grindelia, Gutierrezia, Machaeranthera,* and *Mentzelia.*

Group 3. This group includes many weeds[37] and most crop plants, grains, and grasses, which rarely contain more than 30 ppm of selenium, most of which is associated with plant protein.

Miller and Byers[489] presented a similar classification, but they described in addition a group having a very limited tolerance to selenium, including two species of *Bouteloua.*

This type of classification has many shortcomings, but it suggests a need for caution in generalizing concerning the metabolism of selenium by plants, and study of the literature on selenium strengthens the suggestion.

SELENIUM CONTENT OF PLANTS

A number of factors are involved in determining the selenium content of plants. In the first place, the method of preparing the sample for analysis is important. For instance, the loss of the element from accumulator plants on drying has been well established.[40,41,442] Although losses during the drying of crop plants under mild conditions have not been detected by the usual chemical methods of analysis,[41,178] they have been detected by using radioisotope techniques[24,441,442] and have also been reported at temperatures above those commonly used for drying tissues.[512]

Thrift of the plant[41,509] and weather[94,213] have been cited as causes for variations in the selenium content of plants. However, there are no experimental data showing the effect of these factors on the variations, and there is some question as to what effect, if any, they have.

Different tissues in a plant contain different amounts of selenium, but apparently a number of factors are involved here, and it is impossible accurately to predict from the analysis of one tissue how much selenium will be found in another from the same plant.[41,53,203,247,252,361, 379,509,641,642] Because of its association with proteins and certain amino acids, the element tends to distribute with these in plant tissues.

In general, soils of higher selenium content produce plants of higher selenium content,[489] but, because of the importance of the form of selenium in determining its absorption, the relationship is not strict. Selenium occurring as inorganic selenides or in the elemental form is very insoluble and is not readily absorbed by crop plants.[232,356,509] It would not be important in rendering these plants toxic. It has been suggested that the accumulator plants have the capacity to absorb these forms,[41] and, because of this presumed ability to change the insoluble to a soluble and thus available form, they at one time were referred to as "converter" plants.[40,207,509] Their importance in this solubilization has not, however, been established, and the term is not now commonly used. It has been found that colloidal elemental selenium can be absorbed in very small amounts and solubilized by nonaccumulators.[89,254,603,805] The elemental form, or possibly inorganic selenides, may thus contribute a small fraction of its selenium to a plant, but even in selenium-deficient areas it is doubtful that this would be of much significance.[254] The selenite form of the element is readily available to plants from sand cultures or from nutrient solutions,[89,460,757,759,768] but this is often not true in soils. Some Hawaiian soils of high selenium content produce plants of very low selenium content, presumably because the element occurs as a very insoluble basic ferric selenite.[96,97] Further, soil colloids, again those presumed to contain iron oxides, tightly bind selenite and greatly reduce its availability to plants.[213,247,248,251] Selenates occur in alkaline soils of semiarid areas in high concentration,[96,573,827] where they are quite stable.[18] They are water-soluble, easily leached,[18] and very available to plants.[248,249,509] In areas of excesses of the element, selenates probably contribute the greatest part of the element to plants,[94] although water-soluble organic selenium from decaying plants may contribute significantly in this respect under some conditions.[41]

The distribution of selenium, especially of selenate, throughout the soil profile is of some importance in determining the selenium content of plants. In arid and semiarid areas, the element in its soluble forms can be leached from the upper horizons and redeposited deeper in the profile.[40,41,93,573] Thus, deep-rooted plants may absorb more selenium than shallow-rooted ones. It has been found, however, that soluble selenium deposited even in the second or third foot of the soil profile can contribute significantly to the selenium content of shallow-rooted crops, such as the grasses.[573]

The reduction by sulfur of selenium uptake by plants has been well documented in laboratory experiments.[253,356,357,360,361,695] In the field, Franke and Painter[213] found that sulfur additions had no effect on the selenium content of crops grown on highly seleniferous soils. They concluded that such additions did not have promise as a practical method for reducing selenium uptake by crops, largely because seleniferous soils seemed already to have a high sulfur content in addition to their high selenium content. On the other hand, Allaway,[9] in reviewing sulfur–selenium relationships in plants growing in selenium-deficient areas, suggested that if sulfur fertilization is used it could reduce the selenium content of crops and thus increase the need for supplementation of animal diets with the element. Plants vary in selenium content with their stage of growth.[41,509,569] While in general the selenium content decreases with advancing maturity during the growing season, there are many exceptions to this, and the reasons for the exceptions are not clear.

In the field, plant associations have been found to be somewhat involved in this matter. For instance, grasses growing near accumulator plants have been found to contain more selenium than those growing nearby but not in association with the accumulators.[509] The addition of various proteins and amino acids or plant extracts has increased selenium uptake by plants from culture solutions.[757,759,760] The pH, colloid content, and time have also been found to influence selenium uptake from soils of low selenium content.[247,249,251,768] Undoubtedly, there are additional factors that have some influence in this matter.

TOXICITY OF SELENIUM TO PLANTS

Years before recognition of the fact that selenium is a naturally occurring toxin causing livestock problems, the toxicity of selenium to certain plants under laboratory conditions was known.[27,100,438,729,765] In some of the nonaccumulator species, soluble selenium compounds are injurious to seed germination[438] and growth.[355,357,360,438,459,551,694,695] The characteristic symptom of selenate injury, at least in some cereal crops, is a snow-white chlorosis.[357,765] Although leaves injured by selenite are often greener than normal ones,[355,359] white chlorosis has been reported to occur at very high concentrations of this form of the element.[459] Roots poisoned by selenite may take on a pink color,[27,355,438,729,765] probably because of the precipitation of the colloidal form of the element in the cells. Selenate has not been observed to have this effect.

Walker and Ting[797] have found that selenium reduced the rate of crossing-over in barley. Cytologic observations suggested that the element caused a relaxation of the meiotic chromatin. The question of whether the mechanism by which this occurred would be harmful to plants under some circumstances deserves attention.

The toxicity of the element to plants is influenced by a number of factors. The kind of plant is, of course, very important, the accumulators already mentioned being able to absorb high levels without apparent injury. Some effect of radiation from radium in reducing selenite toxicity has also been recorded.[729] Other factors relate to the uptake of the element, and these have already been discussed.

There are no recorded instances of naturally occurring selenium causing damage to plants in the field. It appears that, in the field, readily available selenium is not provided to plants at a level high enough to cause injury. According to Rosenfeld and Beath,[641] crop plants show no injury until they contain at least 300 ppm of the element, which is usually far in excess of what these plants have been reported to contain even in our most seleniferous areas.

SELENIUM AS A MICRONUTRIENT

The early observation that the primary selenium accumulators grew only where selenium was present in soils and that they could absorb high levels without being damaged prompted the opinion that it might be an essential element in these plants. Indeed, selenite added to sand cultures was found to stimulate the growth of some indicator plants,[729,761,762] strengthening this belief. Further, prior to this time Levine[438] had reported growth stimulation in lupin seedlings as a result of adding selenium dioxide or selenic acid to a distilled-water medium at very low levels. Several others have reported on the stimulation of crop-plant growth by low levels of selenium,[355,602,727] but the data are not impressive and the experiments were not carefully controlled. More recent studies have failed to show any beneficial effect of selenium on the growth of alfalfa or subterranean clover,[75] but work with *Astragalus* species confirms the possibility of its being an essential micronutrient, and the requirement for it is probably very low.[74] Several reviews have covered this question,[18,641,690] and in all cases the need for more work to resolve it is expressed.

CHEMICAL FORMS OF SELENIUM IN PLANTS

Early in his studies of the "alkali disease" syndrome, Franke[207] found the toxic factor (soon after reported to be selenium) in wheat and corn grains to be associated with protein. Then the selenium in some primary indicator plants was found to be largely water-soluble and readily available to plants grown on soils to which it had been added,[38] suggesting its occurrence in some form other than protein. These differences between grains and the indicator plants have been confirmed many times.

Most of the selenium in the grains was found to be firmly bound in the protein and organic in form.[214] On acid hydrolysis of the protein, the selenium was solubilized except for a small amount that remained with the humin, and the substitution of selenium for the sulfur in cystine and methionine was suggested.[580] The probability of such a substitution had been proposed by Cameron[100] as early as 1880. Other studies prior to 1940

Biologic Effects

gave some additional evidence that selenium might occur in place of sulfur in plants.[215,359,581,582] This and some reports on selenium absorption by plants indicated that these two elements were metabolized by similar pathways,[357,358,360,361,551] but on finding that different parts of the same plant had different Se:S ratios, Painter and Franke[579] concluded that there had to be differences.

During some of their early studies of seleniferous plants, Wyoming workers[38] observed that toxicity and offensive odor were related in seleniferous *Astragalus bisulcatus*. They discovered that drying of certain of the primary indicator plants was accompanied by loss of a large amount of selenium,[41] indicating the presence of volatile selenium compounds. They found no such loss on drying grasses or cereal crops. However, *Medicago sativa* plants have been shown to release up to 30% of their selenium in volatile form when dried at 70° C for 48 hr,[24] and the loss of some of the element from grains in long-term storage and in grains heated at 160° C or above for a few hours has been reported.[512]

The first successful attempt to isolate an organic selenium-containing crystalline material was announced in 1940 by Horn and Jones[347] and was described shortly thereafter.[346] The material was believed to be a mixture of the isomorphic compounds cystathionine and Se-cystathionine in a 2:1 ratio. These occurred in free form in hot-water extracts of *Astragalus pectinatus*.

Beath and Eppson[39] studied a number of species of plants and divided them into three classes: (1) those containing largely organic selenium, (2) those containing more than 70% of inorganic (selenate) selenium, and (3) those containing a mixture of organic and inorganic selenium less than 60% of which was in the selenate form.

Smith[709] reported finding selenium concentrated on paper chromatograms of acid hydrolysates of seleniferous protein from wheat or corn grain at locations corresponding to the locations of Se-cystine and Se-methionine. On the other hand, Whitehead *et al.*[824] were unable to confirm this with proteins from grain or cytoplasm of wheat grown on $Na_2{}^{75}SeO_4$.

The work described above was reported prior to 1960, and up to that year the only organic selenium compound identified with reasonable certainty in plants was the Se-cystathionine of Horn and Jones. In 1960, however, Trelease *et al.*[758] reported the isolation of crystalline Se-methylselenocysteine (apparently somewhat contaminated with its sulfur analog) from *Astragalus bisulcatus*. Since then, this compound has been reported in *A. bisulcatus*,[119,548,700,701] *A. crotalaria*,[700,701] *A. canadensis*,[789] *A. succulentus*,[789] *A. vasei*,[789] *A. preussii*,[701] *A. osterhouti*,[461] *A. pattersoni*,[701] *A. sabulosus*,[701] *A. racemosus*,[461,549] *A. pectin-*

atus,[461,462,546] *A. drummondii*,[462] *A. adsurgeons*,[462] *Oonopsis condensata*,[546,700] and *Stanleya pinnata*.[462,701]

In the last decade, a number of other compounds have been identified with reasonable certainty in plants, as follows: Se-methylselenomethionine,[441,605,701,789,791] selenocystathionine[389,461,462,546,604,701,790] and its glutamyl peptide,[547] dimethyl selenide,[441] dimethyl diselenide,[185] γ-L-glutamyl-Se-methylseleno-L-cysteine,[546,548] selenomethionine,[89,570,605] selenite,[89,605] and selenate.[39,89,306] Selenohomocystine[791] and selenocystine and some of its oxides[89,366,605,709,720] have also been reported in plants, but the evidence is less convincing, particularly in the case of selenocystine. Martin and Gerlach[461] also reported the possible occurrence of selenocystine in some *Astragalus* plants, but they suggested the possibility that the peak on which this observation was based may have been the result of a buffer change during the chromatography. Further, Walter et al.[799] caution that diselenide–sulfhydryl and diselenide–selenol interchange reactions occur spontaneously over a wide pH range, and this must be taken into account during investigations of materials containing these mixtures. Particularly in plants having high S : Se ratios, it would be surprising to find selenocystine as such, and half-selenocystine might be involved in a number of combinations with sulfides or other selenides, making its identification difficult.

SELENIUM IN ACCUMULATOR AND NONACCUMULATOR PLANTS

Shrift[690] reviewed the experimental evidence for differences in the metabolism of selenium by the so-called accumulator and nonaccumulator plants. Although Se-methylselenocysteine levels were higher in several accumulator *Astragalus* species,[700,701] other work[789] suggested that what really distinguished the nonaccumulators was the presence in them of Se-methylselenomethionine. This compound occurred only in minute amounts or could not be identified in accumulators. Later, however, he and his co-workers[462] failed to find this amino acid in several nonaccumulator *Astragalus* species and suggested the use of Se-methylselenocysteine and Se-selenocystathionine for distinguishing accumulators from nonaccumulators.

Perhaps it is not surprising that in rather closely related species of plants, sharp differences may not consistently be found, and continued investigations may yield improved methods for differentiating accumulator from nonaccumulator species. In the meantime, the available evidence points to the formation of compounds such as Se-methylselenocysteine as a possible detoxification mechanism for plants that accumulate

Biologic Effects

high levels of selenium, possibly preventing the incorporation of Se-selenocysteine into proteins.[548,700,789]

METABOLIC PATHWAYS

The selenium compounds thus far identified in plants are all analogs of sulfur compounds also found in nature, which suggests similar metabolic pathways for the two elements. Yet many observations indicating differences in their metabolism have been reported. For instance, Nissen and Benson[550] report that the excised roots of several crop plants fed selenate did not form detectable amounts of choline selenate after being treated up to 24 hr, whereas those fed sulfate formed appreciable amounts of choline sulfate. Further, although eel grasses contain flavonoid sulfates, no flavonoid selenates were detected in eel grasses kept for 65 hr in aerated seawater containing $^{75}SeO_4^{-2}$; and although 6-sulfo-α-D-quinovopyranosyl-(I→I')-2',3'-diacyl-D-glycerol occurs in high concentrations in plant photosynthetic material, its selenium analog could not be detected in a number of plants for which the selenate uptake periods lasted several days. The apparent absence of cystathionine from plant extracts showing the presence of selenocystathionine,[462] the absence of selenoglutathione from plants that synthesize glutathione itself,[701] and the many quantitative differences in selenium and sulfur compounds in plants further attest to some differences in the metabolism of the two elements. In brief, the metabolism of selenium in any plant apparently cannot be derived from a study of its sulfur metabolism.

Unfortunately, our knowledge of the metabolic pathways for selenium in plants is very limited. Failing to detect 3'-phosphoadenosine-5'-phosphoselenate in plants grown in selenate, Nissen and Benson[550] suggested the following for the reduction of selenate to selenite:

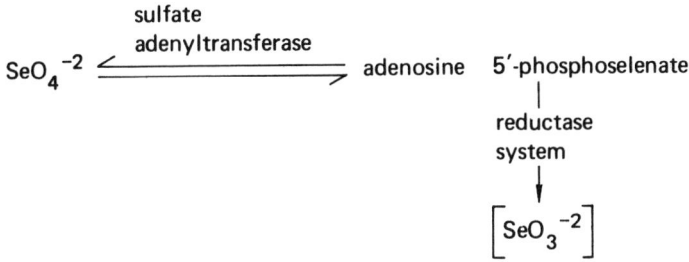

In regions of alkaline soils, selenate is probably the form in which the element is absorbed by plants.[18] Selenite has been found to be more readily absorbed and metabolized from culture solution than is selenate,[89,701] and Butler and Peterson[89] suggest that the reduction to selenite may be the rate-limiting step in the metabolism of selenate.

Virupaksha et al.[791] proposed the following for the biogenesis of selenohomocystine in *A. crotalaria*:

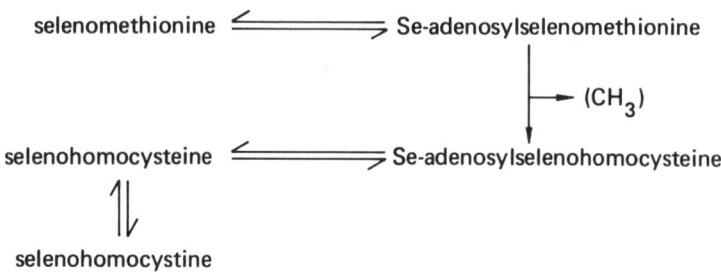

Lewis et al.[441] reported a crude enzyme preparation from *Brassica oleracea* var. *capitata* that cleaved Se-methylselenomethionine into homoserine and dimethyl selenide, a compound that had been identified in this plant. The preparation also cleaved S-methylmethionine into dimethyl sulfide and homoserine. Dimethyl selenide could not be identified in *A. bisulcatus*, but dimethyl diselenide could. Froom,[220] however, reported the presence of the dimethyl selenide in this plant at an earlier date.

Even the insoluble elemental selenium in colloidal form can be absorbed by *Spirodela oligorrhiza* from culture solution.[89] Labeling patterns for the various selenium compounds formed were the same regardless of whether selenite, selenate, or elemental selenium was fed the plants, which suggests a common pathway for the metabolism of these three forms of the element in this plant.

A much more detailed review of the metabolism of selenium by plants was recently prepared by Shrift.[692]

Microorganisms

Research into the metabolism of selenium in microorganisms furnishes us with evidence of fundamental biochemical concepts that may elucidate the principal role of selenium as a nutrient and toxicant in all living forms.

TRANSPORT ANTAGONISMS BETWEEN SULFUR AND SELENIUM COMPOUNDS

Ample evidence supports the concept that chemically similar selenium and sulfur compounds can compete with one another for transport across

Biologic Effects

the cell membrane of a variety of microorganisms.[691] For example, sulfate transport in *Penicillium chrysogenum* was inhibited by selenate, and the selenate was shown to enter the mycelium via a sulfur-regulated permease thought to be the sulfate permease.[842] A similar competitive antagonism was observed in *Chlorella vulgaris* in respect to the uptake of selenate vs. sulfate or selenomethionine vs. methionine.[694,695]

However, there are a number of cases in which the nature of the metabolic interrelationship between selenium and sulfur is not as obvious as in those referred to here. As expected, sulfate could reverse the toxic effect of selenate on yeast growth,[193] but a similar reversal also could be obtained with methionine.[194] Moreover, there appeared to be a difference in response between species: methionine had no activity in reversing the toxic effects of selenate on *Escherichia coli*, but cysteine and glutathione did.[192] In fact, methionine has been reported to enhance the toxicity of selenite to *E. coli*.[650] In this instance, exogenous methionine was thought to suppress the conversion of selenite to selenomethionine, which was considered a detoxification product. Experiments with yeast point to the possibility of metabolic antagonisms between selenite and chemically dissimilar anions; the inhibition of yeast respiration caused by selenite could be reduced by arsenite, arsenate, or phosphate.[61] More work is required to clarify these relationships. The toxicity or availability of selenium to microorganisms can clearly be influenced by the nature of the other compounds present in the growth medium.

REDUCTION AND OXIDATION OF SELENIUM COMPOUNDS

Extracts of *Micrococcus lactilyticus* were shown to utilize molecular hydrogen to reduce a wide variety of oxyanions, including selenite.[833] The reduction of selenite consisted of two steps: a rapid reduction of selenite to elemental selenium followed by slower reduction of the colloidal selenium to selenide. Chemically prepared suspensions of colloidal selenium were also found to be reduced to selenide. Selenate, on the other hand, was not reduced. Selenite reduction by *Salmonella heidelberg* was also shown to be a two-stage reaction, but in this case the final product was elemental selenium rather than selenide, and the intermediate reduction was trapped and identified as the divalent positive selenium ion.[481] The authors suggested that the tolerance of *Salmonella* to selenite is due to the conversion of selenite to the insoluble and nontoxic elemental selenium.

The reduction of selenite by cell-free preparations from yeast has been investigated in detail by Nickerson and Falcone.[545] These workers found that dialyzed enzyme preparations obtained from bakers' yeast or *Can-*

dida albicans could reduce selenite to elemental selenium if they added back to the system either the dialyzable substances or a boiled undialyzed extract. Extraction of the boiled yeast extract with *n*-hexane removed its ability to restore selenite-reducing capacity in dialyzed enzyme preparations. Menadione or thiadione could replace the substances extracted by *n*-hexane. The selenite appeared to be bound to protein through vicinal thiol groups. The authors postulated the following pathway of electron flow to account for the reduction of selenite:

$$TPN \longrightarrow Flavin \longrightarrow \underset{\underset{Protein}{S\ \ CH_3}}{O=\bigcirc=O} \longrightarrow \underset{\underset{Protein}{S\ \ S}}{O=Se} \quad Se^{+4} \rightsquigarrow Se^0$$

A role for flavin in selenite reduction was also suggested by the work of Tilton et al.,[753] who found that flavin-adenine dinucleotide (FAD) was required for the maximum reduction of selenite to elemental selenium by cell-free extracts of *Streptococcus faecalis* or *Streptococcus faccium*. These workers also found that the reduction of selenite could be inhibited by a variety of sulfhydryl blocking agents, which at least is consistent with the idea that sulfhydryl groups are present at the active site of the selenite-reducing enzyme. The ability of microorganisms to reduce soluble selenium compounds to the insoluble elemental state assumes practical importance when it is realized that such so-called selenoreductase activity is higher in microbial strains that have adapted to high-selenium conditions.

In contrast to the well-documented reductive pathways of selenium metabolism in microbes, there is relatively little evidence of existence of oxidative pathways of selenium metabolism in microorganisms.[693] This fact could have important consequences for the environmental cycling of selenium.

BIOSYNTHESIS AND METABOLISM OF SELENO-AMINO ACIDS

Several investigators have used chromatographic identification as evidence for the conversion of inorganic selenium into seleno-amino acids by microorganisms. *Escherichia coli* grown on a sulfur-deficient medium containing radioselenite incorporated trace quantities of selenium into compounds that had chromatographic properties similar to those of

Biologic Effects

synthetic selenomethionine.[767] The presence of selenocystine could not be demonstrated. Similar results were reported for a selenium-tolerant substrain of *E. coli* when radioselenate was used as the source of selenium.[353] Blau[56] grew yeast on a low-sulfur medium containing ^{75}Se-selenite and isolated a material by ion-exchange chromatography that was considered to be selenomethionine. The selenomethionine produced under his conditions could be incorporated into proteins and had biologic properties much like those of methionine. Weiss *et al.*[811] found compounds having R_f values corresponding to selenocystine in *E. coli, Proteus vulgaris*, and *Salmonella thompson*, but selenomethionine was detected only in *E. coli*. Studies with cultures of mixed rumen bacteria have yielded conflicting results concerning the formation of selenomethionine from inorganic selenium by rumen microorganisms. Hidiroglou *et al.*[334] found that rumen bacteria could incorporate selenite into the microbial protein. Characterization of the selenium compounds in the rumen bacteria protein hydrolysates by chromatography suggested the presence of selenomethionine. Paulson *et al.*[596] also found that inorganic selenium could be incorporated into the trichloroacetic-acid-insoluble fraction of rumen fluid. However, when the trichloroacetic-acid-insoluble fraction was dialyzed against reduced glutathione, most of the selenium could be removed and therefore must have been rather loosely bound. These workers concluded that very little or none of the inorganic selenium added to the rumen fluid was incorporated into selenomethionine. The sole use of chromatographic techniques has led to incorrect identification of selenium compounds of biologic interest.[669]

Several investigators have shown that selenomethionine can be utilized effectively as a substitute for methionine in a number of *in vitro* enzymatic reactions. Mudd and Cantoni[515] found that selenomethionine could substitute for methionine in the reaction catalyzed by yeast methionine-activating enzyme. Furthermore, the product thereby formed, Se-adenosyl-selenomethionine, could serve effectively as a methyl donor in the methylation of guanidoacetic acid to form creatine. McConnell and coworkers demonstrated that selenomethionine can participate in all the known reactions of methionine during polypeptide chain initiation and synthesis in *E. coli*,[340,470] but the relation of these findings to the overall protein biosynthetic pathway is not clear.

It is generally agreed that exogenously supplied selenomethione can be incorporated into the proteins of microorganisms. There is, however, some controversy regarding the biologic consequences of such incorporation. In an early study, Cowie and Cohen[132] claimed that selenomethionine could completely replace methionine for the normal exponential growth of a methionine-requiring mutant of *E. coli*. Wu and Wachsman,[839] however, showed that selenomethionine only partly satisfied the methionine requirement of methionineless strains of *E. coli* WWU or *Bacillus megaterium* KM. Coch and Greene[125] found marked strain differences in the toxic effects of selenomethionine on *E. coli*. Selenomethionine was relatively nontoxic to the growth of *E. coli* 26 at concentrations as high as 0.01 M as long as cysteine was added to the growth medium, but the growth of *E. coli* K12 was markedly inhibited by 10^{-4} M selenomethionine, regardless of whether cysteine was present. The catalytic activity of *E. coli* 26 β-galactosidase with 70%–75% of its methionine residues replaced by selenomethionine was found to be the same as that of the unmodified enzyme. These results are similar to those of Huber and Criddle,[352] who found that selenomethionine substitution of about 80 of the 150 methionine residues of the β-galactosidase from a selenium-tolerant substrain of *E. coli* K12 grown on 0.01 M selenate had no effect on the catalytic parameters K_m and V_{max} of the enzyme. These workers noted, however, that the stability of the selenium β-galactosidase when subjected to heat and urea was decreased. Selenomethionine was shown to cause a severe inhibition of both bulk protein and β-galactosidase synthesis in *E. coli* 26 if amino acids other than methionine were limiting in the growth medium.[125]

METHYLATION OF INORGANIC SELENIUM COMPOUNDS

Several species of molds, especially *Scopulariopsis brevicaulis* (*Penicillium brevicaul*), methylate oxyanions of selenium, tellurium, or arsenic to give volatile compounds with a characteristic garliclike odor.[113] Thus, Na_2SeO_3, K_2TeO_3, and As_2O_3 yield $(CH_3)_2Se$, $(CH_3)_2Te$, and $(CH_3)_3As$, respectively. The mechanism postulated for this conversion is the following:

$$H_2SeO_3 \rightarrow H^+ + :Se(O^-)(=O)(-OH) \xrightarrow{CH_3^+} CH_3Se(=O)(=O)(-OH) \xrightarrow{\text{ionization \& reduction}} CH_3Se:(O^-)(=O)$$

ION   METHANESELENONIC ACID   ION OF METHANESELENINIC ACID

$$\xrightarrow{CH_3^+} (CH_3)_2Se(=O)(=O) \xrightarrow{\text{reduction}} (CH_3)_2Se:$$

DIMETHYL SELENONE   DIMETHYL SELENIDE

Biologic Effects

Later work established that the source of methyl groups for this reaction was methionine.[161]

In 1972, Fleming and Alexander[205] found that microorganisms isolated from raw sewage can produce dimethylselenide from inorganic selenium compounds. The ecologic consequences of such a process are not known, but it should be pointed out that the methylation of inorganic selenium oxyanion salts probably does not pose the same kind of ecologic threat as the methylation of inorganic mercury compounds, since dimethyl selenide itself is much less toxic to mammals than inorganic selenium oxyanion salts.[474] However, Pařízek et al.[589] described a remarkable synergistic toxicity between dimethylselenide and traces of inorganic mercury salts.

ROLE FOR SELENIUM IN FORMATE DEHYDROGENASE

By using a highly purified glucose-minimal salts culture medium, Pinsent[610] was able to show that traces of selenite and molybdate, as well as iron, were needed for the production of formic dehydrogenase in *Escherichia coli*. These factors were effective only if added during cell growth and had no effect if added to washed cell suspensions. Several years later, Fukuyama and Ordal[226] found that iron-deficient cells of *E. coli* exhibited formic dehydrogenase activity only when adequate amounts of selenium and molybdenum were present in the growth medium.

In 1971, Lester and DeMoss[427] demonstrated that selenite was required to form the enzyme system that permits formate to serve as an effective electron donor for nitrate reduction in anaerobically grown *E. coli*. As pointed out by the authors, anaerobic electron transport in *E. coli* is being studied intensively as a model of the mechanisms of synthesis, assembly, and regulation of membrane-bound enzyme systems. Since the effects of selenite and molybdate on formate dehydrogenase could be blocked by chloramphenicol, it appeared that protein biosynthesis was required for these effects. The selenite–molybdate requirement was quite specific for enzymes of formate and nitrate metabolism; selenium and molybdenum had no effect on the level of several other dehydrogenase and oxidase systems.[182] DL-Selenocystine was about as effective as selenite in stimulating the formation of formic dehydrogenase, whereas DL-selenomethionine was only 1% as effective. The authors speculated that selenium could be an integral part of the enzyme formate dehydrogenase and could have a catalytic role either as selenocysteine or as nonheme iron selenide. Shum and Murphy[702] found that ^{75}Se incorporated by *E. coli* migrated with formic dehydrogenase activity through a sucrose density gradient.

ADAPTATION TO SELENIUM

Shrift and Kelly[696] found that *E. coli* K12 exposed to 2×10^{-4} M of selenate attained growth rates similar to controls after a lag period of 24-48 hr. These selenium-tolerant *E. coli* apparently became resistant to selenate: the cells would grow immediately when placed in a new high-selenate medium, whereas *E. coli* not previously exposed to selenate grew only after a long lag phase. This selenium-tolerant substrain maintained its resistance to selenium after nine transfers, which suggested that the adaptation was stable. No mechanism was proposed. In a later paper, Huber *et al.*[353] reported somewhat different growth characteristics in a similar selenium-resistant substrain of *E. coli* K12. The much shorter lag times and exponential growth observed by Shrift and Kelly were considered to be due to a more extensive sulfur contamination in the chemicals used to prepare their growth media. Shrift *et al.*[697,698] also described a permanent adaptation of *Chlorella vulgaris* to selenomethionine. The resistance in this case appeared to be due to a decreased permeability of the algal cell to either methionine or its selenium analog.[699] Springer and Huber[723] noted a decreased uptake of selenate in two selenate-tolerant strains of *E. coli* as compared with uptake in wild *E. coli*. Kovalskii and Ermakov[404] showed that microorganisms taken from geologic zones high in selenium were less susceptible to the toxic effects of the element than were microorganisms taken from soils low in selenium. One mechanism for such resistance could be increased levels of selenoreductase.[405] This possibility is suggested by the fact that strains of *Bacillus megaterium* taken from seleniferous soils had higher levels of this enzyme than strains taken from soils low in selenium.[428]

NUTRITIONAL, PROPHYLACTIC, AND THERAPEUTIC USES

Selenium is an essential nutrient for chicks and quail,[99,541,542,594,749,750] rats,[480,666,667,820] and sheep.[80,320,520,524] There is strong support for its essentiality in swine[177,190,483,576,599,755,785] and cattle,[319,320,333,336,455,522,525,535,555] and there is an apparent need for it in squirrel monkeys.[528] Other species in low-selenium areas may have low concentrations of selenium in their feedstuffs and tissues[81,133,231,271,327,375,376,409,592,616,647,671] and might benefit from prophylactic or therapeutic administration of selenium salts or a mixture of selenium salts and α-tocopherol.

Selenium deficiency has been induced in rats, sheep, and squirrel monkeys by use of low-selenium feeds supplemented with 60 μg of α-

Biologic Effects

tocopherol per gram of feed or 720 IU of vitamin E per ewe per week.[480,528] Deficiency lesions were prevented or reversed by the addition of sodium selenite or selenate to the feed (100 ng/g) or by parenteral injection (1 mg/50 kg body weight).[480,660] Neither parenteral injection nor additional dietary supplementation with α-tocopherol prevented or reversed deficiency lesions.

Selenium Deficiency and Its Control with Selenium Salts and α-Tocopherol (Vitamin E)

Since 1960, selenium salts and mixtures of selenium salts and vitamin E have been widely used in selenium-deficient areas throughout the world.[272,309,319,521,558,718] They have been administered as prophylaxis to pregnant ewes, cows, and sows; to neonates of these species; and to rapidly growing lambs, calves, pigs, chicks, and poults.[7,8,186,272,409] These preparations have also been widely used in dogs and horses to correct clinical signs of musculoskeletal weakness, lameness, dermatitis, infertility, and abnormal hair coat.[137,319,840]

The usual route of administration has been parenteral. However, in Australia and New Zealand sodium selenate has been added to mineral salts or fertilizer or used as a supplement in mixed feeds.[7,8,320] Permission for supplementation of mixed feeds for young poultry and pigs was granted in Canada on September 6, 1973, and in the United States on February 7, 1974, by the U.S. Food and Drug Administration.

Selenium supplementation has prevented nutritional myopathy (white muscle disease) in sheep, swine, and cattle;[272,524,660] hepatic necrosis (hepatosis dietetica), myocardial necrosis and hemorrhage (dietetic microangiopathy or mulberry heart disease), gastroesophageal ulceration, and "Herztod" in swine;[272,785] steatitis and exudative diathesis in chicks and poults;[269] lameness in dogs, horses, and breeding bulls;[30,31] and poor growth and reproduction in sheep and rats.[840]

Parenteral and nutritional use of selenium salts in biology was initiated by the 1957 discovery that selenium was the third factor, along with cystine and vitamin E, that prevented massive liver necrosis in rats fed a specific torula yeast ration.[666]

In 1957, Muth and associates reported that sodium selenite but not α-tocopherol prevented nutritional myopathy in sheep injected with 60 mg of α-tocopherol per kg of body weight.[524] Regardless of whether ewes were treated with α-tocopherol, parenteral injection of 1 mg of selenium as selenite per 50 kg of body weight, once a month, prevented myopathy in their lambs. Lambs born to ewes that were not treated with selenite developed myopathy.[526,527]

Domestic Animals

NUTRITIONAL MYOPATHY OF SHEEP AND CATTLE

Nutritional myopathy (white muscle disease) became prevalent in ruminants in the northeastern and northwestern parts of the United States and in Australia, New Zealand, and northern Europe after World War II.[522] The appearance of the syndrome in two Oregon counties (about 1950 and 1968) was associated with changes in methods of forage and sheep production.* Low-yield hay production and grass foraging supplemented with grain and protein concentrates were replaced with intensified high-yield grass production and the rearing and marketing of lambs and ewes with minimal grain feeding.[186,524]

In these Oregon counties, the soils were more deficient in sulfur than in any other plant nutrient. Thus, the change to high-quality, high-yield grasses was associated with heavy fertilization with sulfate fertilizers.[72] Hays grown in these areas now often contain less than 20 ng of selenium per gram (dry-weight basis).*

Selenium deficiency occurred in sheep and lambs after the initiation of more productive methods.[524] The reasons for the development of selenium deficiency have not been determined. Perhaps there was a lack of available selenium in the soil relative to the increased yield of forage and protein, or sulfate fertilizer may have prevented utilization of selenium,[817] or cessation of grain feeding may have removed a major source of selenium for growing lambs.

Allaway reported that fertilization of heavily cropped fields with a few ounces of selenium salts per acre prevented selenium-deficient forage.[7] This practice has been used in Australia and New Zealand with satisfactory results, although mistakes in application have produced selenosis in some areas.[229,230,688]

Trinder et al.[763] reported on the effect of selenium in preventing retained placenta in dairy cattle. Several other authors have reported on the distribution of selenium and radiotocopherol in pregnant ewes and fetal lambs.[87,837]

Since 1955, Muth and associates have produced nutritional myopathy in lambs from an experimental flock of ewes fed selenium-deficient alfalfa hays from selected fields in northeastern Oregon.[524,526,527,529,600,817,818,821-823] These hays contained 13–23 ng of selenium per gram of hay (dry weight), the amount depending on the location of the field and the year. The incidence of nutritional myopathy in lambs from this flock was

*J. R. Harr, unpublished observations.

not exactly related to the concentration of selenium in the hay. Hays from certain locations or grown in certain years were more effective in producing myopathy than hays with somewhat lower concentrations of selenium but grown at a different time or place. Causes of these variations are unknown. They may involve lipid components of the forages and rumen activity.

Lesions of nutritional myopathy in sheep and cattle are primary calcification and degeneration of skeletal muscle and myocardium.[63] The left ventricular wall and interventricular septum are usually affected, whereas the auricles, right ventricle, and apex are spared. The more active skeletal muscles, especially the abductors and the longissimus dorsi, are affected.

Lesions of myocardial nutritional myopathy have been observed in lambs aborted during the last month of gestation by ewes in the low-selenium experimental flock at Oregon State University. The young may be born edematous with passive congestion, labored breathing, and weak, irregular pulse.

Both myocardial and skeletal muscles of the neonate may be affected. The skeletal muscles affected are those with the greatest work requirement. In the neonate these are the abductors of the thigh; in older animals, the longissimus dorsi and triceps are affected. Mortality may be 65% during the first 10 days of life. Oral or parenteral administration of 1 mg of selenium as selenite will produce remission of clinical signs within a few hours.

Clinical signs of selenium-deficiency nutritional myopathy may occur or reoccur during the initial phase of rapid growth from 3 to 8 weeks after birth.[320] The predominant signs are skeletal weakness, especially of the semitendinosus–semimembranosus group of muscles and the longissimus dorsi, and slow growth.[59] The connective tissue around the distal portions of the semimembranosus and semitendinosus muscles may contain a heavily proteinaceous exudate.* The concentration of lactic dehydrogenase; 5'-nucleotidase; and glutamic, oxalic, and pyruvic transaminases in serum is increased.[79,600,821,822] Lysosome and lysosomal enzyme changes have been reported.[71,79,418,823] Mortality may be 35%. Parenteral administration of 1 mg of selenium as selenite per 50 kg of body weight will reverse clinical signs and both the morphologic and biochemical lesions.[524,600,660,823]

In low-selenium areas, mature sheep and cattle that grow poorly or are unthrifty or have lowered reproductive ability or muscular weakness often improve clinically after administration of 1 mg of selenium per 50

---

*J. R. Harr, unpublished observations.

kg of body weight.[319,524,660,840] Annual repetition of selenium administration or two or three injections during gestation generally prevent reoccurrence of the signs of selenium deficiency in ruminants in low-selenium regions.[59,660] Ruminal pellets that contain elemental selenium are used successfully in Australia to provide needed dietary selenium.[309,410]

The occurrence and initial lesions of nutritional myopathy may differ with differences in climate and in haying methods. In the United States, Australia, and New Zealand, selenium salts are considered more effective than tocopherol in preventing nutritional myopathy.[62,228,320,524,660] Forages are heavily fertilized, the summers are hot and dry, and hay is readily made and easily cured or overcured. The initial histologic lesions of nutritional myopathy in these regions are microscopic deposits of calcium midway between the Z-bands of the sarcomere.[519 (p. 225)] These deposits are less than a micrometer in diameter and may coalesce to form grayish white plaques of calcium that extend along the muscle fibers.

In northern Europe, α-tocopherol has been considered the most effective component of the selenium–vitamin E mixture in preventing nutritional myopathy.[558,718] In these countries, nutritional myopathy is associated with poor curing of lightly fertilized native grasses and with an alteration of the unsaturated fat components in the cured hays. In these conditions, the primary morphologic lesion of the sarcomere may be degeneration, and calcium deposition is secondary and delayed, or absent.

SUDDEN DEATH, HEPATIC NECROSIS, ARTERIOLAR DEGENERATION, AND SKELETAL AND CARDIAC MYOPATHY IN SWINE

Trapp and associates and others reported that a mixture of selenium and vitamin E prevented a deficiency syndrome of swine in Michigan that was characterized by sudden death of feeder pigs, hepatic necrosis, icterus, edema of the mesentery, fibrinoid degeneration and microangiopathy of the media or the arterioles, and skeletal and cardiac myopathy.[190,270,755,785,801] Others have reported on ultrastructural and histochemical changes in selenium-deficiency myopathy[739] and hepatic necrosis[456,483] in pigs.

Van Vleet and associates in Indiana described the effect of the selenium–vitamin E mixture in growing swine raised on premises where hepatosis dietetica and mulberry heart disease were common.[785,786] They concluded that supplementation of pregnant sows and baby pigs was necessary for profitable swine husbandry in these areas.

Biologic Effects

The concentration of selenium in feeds associated with this deficiency syndrome in field conditions was 20–60 ng/g.[785,786] Mortality of young pigs reared under these conditions was 10%. α-Tocopherol did not control the hepatic, muscular, or vascular lesions. Effects of prophylactic administration of selenium–vitamin E mixtures on tissue and blood composition were reported by Ewan and associates.[189,190]

Twenty pregnant sows on a farm with a history of selenium deficiency, hepatosis dietetica, mulberry heart disease, and feeds that contained little selenium were inoculated intermuscularly with 5 mg of selenium as selenite (about 1 mg/25 kg),[786] and 40 other sows from the same farm were given placebo injections. Of 538 3-day-old pigs from these 60 sows, 341 were inoculated with a mixture of selenium and vitamin E in doses of 60 ng of selenium as selenite per kilogram of body weight. The other 197 pigs were given placebo injections.

The incidence of stillborn pigs from the 40 placebo-treated sows was 9%, compared with 3% in the 20 selenium-treated sows. Neonatal deaths were 6.2% and 3.4%, respectively. Deaths to 6 months of age (about 225 lb) were 17% and 11%, respectively. One of the 341 pigs that received the selenium–vitamin E mixture died of selenium-deficiency disease; 14 of the 197 pigs from the placebo-treated sows died of hepatosis dietetica or mulberry heart disease. Treated pigs (i.e., treated with the mixture of selenium and vitamin E) from selenium-treated sows were compared with treated pigs from placebo-treated sows; differences were minor.

The selenium concentration in the liver of stillborn pigs from selenium-inoculated sows (5 mg per sow) was 177 ng/g compared with 88 ng/g in pigs stillborn to placebo-inoculated sows.[786] The liver of pigs that died of hepatosis dietetica or mulberry heart disease contained 60–180 ng of selenium per gram. The concentrations of selenium in the kidneys and liver of the selenium-inoculated pigs were 12% and 21% greater, respectively, than in the noninoculated pigs. However, the concentration of selenium in the muscle and fat of the inoculated pigs was 14%–20% less than in the noninoculated pigs. The ratio of the concentration of selenium in the liver to the concentration in the kidneys was 0.12 in the noninoculated pigs and 0.11 in the inoculated pigs. These ratios (0.12 and 0.11) are similar to those found in the normal to selenium-deficient controls used by Herigstad and associates in studying dietary selenosis in swine.[329] The ratio is about one-third of the ratio found in pigs fed 100 ng of selenium as selenite per gram of feed.

Experimentally produced lesions of selenium and vitamin E deficiency in eight weanling female pigs were increased concentrations of plasma transaminases, creatine phosphokinase, α-hydroxybutyric acid dehy-

TABLE 5-1 Selenium Content of Tissue from 14 Pigs Fed Selenium–Vitamin E Deficient Feeds, Selenium-Supplemented Feeds, and Commercial Feeds[646]

	Selenium Content of Tissue (ng/g—wet weight)	
Feed	Liver	Psoas Muscle
Basal ration (20 ng Se/g)	33	25
Basal ration plus Na_2SeO_3 (300 ng Se/g)	400	65
Commercial ration	265	83

drogenase, isocitric dehydrogenase, lactic dehydrogenase, and selective destruction of Type I skeletal muscle fibers.[646] There was also a decrease of phosphorylase activity in Type II fibers.

Plasma enzymatic activity was increased in pigs that had liver and muscle concentrations of selenium of 20–40 ng/g. Pigs in the control groups had no plasma enzyme activity and had selenium concentrations of 60–100 ng/g of muscle and 230–380 ng/g of liver.

Muscle lesions appear to result from primary fiber damage rather than from vascular lesions, pulmonary edema, fibrinoid degeneration, or neuroangiopathy.[646] This observation is similar to that of Muth about the development of myopathy in sheep.[519]

Addition of selenium to the basal ration as sodium selenite was less effective in increasing the concentration of selenium in muscle than maintaining the pigs on a commercial feed (65 vs. 85 ng/g). However, the concentration of selenium in the liver of selenite-treated pigs was greater than in commercially fed pigs (400 vs. 265 ng/g). Others have reported that natural forms of selenium produce greater concentrations of selenium in muscle than does selenite (Table 5-1).[407]

In a series of experiments by several investigators, weanling pigs fed semipurified diets based on torula yeast as the protein source and adequate amounts of the sulfur amino acids developed liver necrosis and died.[177] Lesions and mortality were prevented by addition of either selenium or vitamin E to the feed, but not by addition of cystine.[599] Pigs fed natural feeds that had been treated to reduce the content of α-tocopherol also developed myopathy. In these pigs the diagnosis was based on an increase in the concentration of glutamic–oxaloacetic transaminase in plasma.

Herigstad et al.[329] in their series on selenosis had four pigs on selenium-deficient rations. Two of these four died of selenium–vitamin E deficiency. Signs and lesions were distress, ataxia, sudden death, hepatic swelling, necrosis and hemorrhage, irregular Glisson's capsule, hemorrhagic ileitis, and ecchymotic hemorrhages. The concentration of selenium

Biologic Effects

in the liver was 110 and 90 ng/g, and in the kidneys 370 and 430 ng/g.

EXUDATIVE DIATHESIS, ENCEPHALOMALACIA, AND MYOPATHY IN CHICKS AND TURKEY POULTS

Patterson et al.[594] demonstrated shortly after the initial reports by Schwarz and Foltz[666] that selenium salts prevented exudative diathesis in chicks fed torula feeds that were low in vitamin E. This report, along with that of Schwarz and Foltz,[666] led to investigations of other species.[666] Subsequent work produced exudative diathesis in both Japanese quail and chicks fed synthetic (amino acid) diets that contained little selenium but high levels of α-tocopherol.[616] Supplementation of these feeds with selenium salts prevented the deficiency.

The syndrome produced in chicks by the synthetic diets supplemented with α-tocopherol included exudative diathesis, poor growth, poor feathering, and fibrotic degeneration of the pancreas.[616] Death usually followed decreased absorption of lipids, decrease in production of enzymes, and failure of fat digestion.[749] Under these conditions, bile production decreased, and there was a decrease in the concentration of bile and monoglycerides in the intestinal lumen. The formation of lipid–bile salt micelle was reduced, and α-tocopherol absorption was decreased.

Addition of fatty acids, monoglycerides, and bile salts to the basal synthetic feed improved absorption of vitamin E. This regimen prevented exudative diathesis but not degeneration of the pancreas.[616,749] Addition of both high levels of vitamin E and 10 ng of selenium as selenite per gram of feed prevented the pancreatic degeneration. When concentration of the vitamin E in the feed was normal (10–15 IU/kg of feed), 20–40 ng of selenium per gram of feed was necessary to prevent pancreatic degeneration.

Feeds low in both sulfur amino acids and selenium produced myopathy of the active pectoral muscles. Lesions were grayish white striations through the muscle.[99] Supplementation with cystine or vitamin E prevents the myopathy. But supplementation with selenium as selenite was only partly effective in preventing it.[541]

The pathology of selenium deficiency in the chick was reported by Gries and Scott.[269] Effects include poor growth and have been prevented by addition of selenium to the drinking water.[613] Experimentally, 50–60 ng of selenium per gram of feed is needed by the chick to prevent exudative diathesis, the amount depending on the type of feed and the amount of vitamin E.[465,552,672,747] Field cases of exudative diathesis have been reported in the United States[542,594] and in New Zealand.[320,647]

Selenium deficiency in turkey poults produces a mild form of exudative diathesis[133] and more characteristically degeneration of the muscle of the gizzard. The pectoral muscles are affected in 25% of the birds. Some birds have myocardial failure.[671] These lesions occur in poults fed a diet deficient in vitamin E and can be prevented by addition of 80–280 ng of selenium as selenite per gram of feed. Methionine and cystine and additional α-tocopherol do not prevent either the signs or the lesions.

Lesions of exudative diathesis are similar to the exudation observed in connective tissue surrounding the digital portion of the semitendinosus and semimembranosus muscles of lambs with nutritional (selenium) myopathy. Nutritional (selenium) myopathy in chicks and poults is a degeneration of muscle fibers with perivascular infiltration and proliferation of histocytes and granulocytic leucocytes.[269] This lesion is similar to those in sheep and pigs. Encephalomalacia of chicks and reproductive failure of hen turkeys are associated with a deficiency of vitamin E and are not affected by the amount of selenium in the feed.[647]

EFFECTS IN OTHER SPECIES

Clinicians in areas where selenium deficiency occurs observed that breeding bulls, rams, dogs, and horses develop nonspecific lameness and muscular tenderness that may respond to selenium–vitamin E therapy.[319,320] Commercial mixtures of these compounds are approved for parenteral use in dogs and horses. They are recommended for muscular weakness, lameness, and tenderness. Gabbedy and Richards[231] reported selenium-deficient myopathy in a foal.

Mixtures of selenium and tocopherol are commercially available for use in several species. They are prepared for use in alleviating and controlling pain and lameness associated with some arthropathies. The manufacturer states that the mixtures are for treating symptoms rather than etiology.

These mixtures are also claimed to be effective therapy in some cases of idiopathic dermatitis. The efficacy of these products in this type of condition may be related to the possible action of selenium as an antiinflammatory agent. Spallholz et al.[719] have reported the effect of supplementation of diets with selenite selenium on the immunologic responses of mice.

Clinical recommendations are made from time to time as to the effectiveness of these mixtures. The disorders to which the recommendations relate include lameness, synovitis, hepatic degeneration, infertility, dermatitis, poor growth, and unthriftiness.

Biologic Effects

Laboratory Animals

RATS

Attempts to develop uncomplicated selenium deficiency in laboratory rats were unsuccessful until 60 ng of α-tocopherol per gram was added to low-selenium (18 ng/g) torula yeast feed.[68,345,363] Burk et al.[84] postulated a threshold level for adequate dietary selenium of about 10 ng/g. Rats from Oregon State University's brown, cotton, and Wistar colonies maintained for one to three generations on this feed grew slowly, had poor hair coats, and were sterile. The tactile hairs were not affected. Further work demonstrated that the second litter born to the Wistar or brown rats maintained on the low-selenium feed were selenium-deficient.[314] Deficiency was more difficult to produce in the cotton rats than in the others. Either the third generation or the second litter from the second generation was maintained on the low-selenium regimen until deficiency developed.

In other experiments, selenium-depleted rats that were not supplemented with additional selenium grew slowly, had poor hair coats, and were sterile.[68,840] Vascularization of the subcutis and dermis was incomplete, and the eyes contained cataracts.[724] The tactile hairs (nourished from cavernous blood sinuses) were not affected. Germinal epithelial cells of the skin and of the endothelial cells of the capillaries and small arteries contained fewer stainable RNA or sulfhydryl groups than did those of rats on the same feed supplement with 100 ng of selenium as selenite per gram.[724] Addition of selenium as selenite or selenate to the feed of 60-day-old, selenium-deficient rats at a rate of 10–100 ng/g resulted in complete reversal of signs within 30–90 days. Similar results were obtained by other authors.[345,363]

In recent work by Johnston,[380] selenium-deficient, tocopherol-supplemented brown rats fed torula yeast feeds ate 50%–60% more feed than littermates fed the same feed with the addition of 2 μg of selenium as selenite per gram of feed (776 vs. 1,198 g in 17 weeks). Feed efficiency was 75% greater in the rats fed the selenium-supplemented diet than it was in the rats fed the basal diet (4.5–5.4 g of feed per gram of weight gained compared with 7.9–8.8 of feed per gram of weight gained).

The concentrations of selenium in the liver, muscle, and kidneys of selenium-deficient rats were, respectively, 0.4, 1, and 2 μg/g (dry weight).[314,345] Addition of selenium as selenite to the feed of these rats at a rate of 100 ng/g increased selenium concentration in the liver to 2 μg/g. Addition of selenium to these deficient feeds at rates of 500 and 2,500 ng of selenium per gram (five and 25 times the 100-ng/g rate of supplementation) increased the concentrations of selenium in liver to 5.6 and 7.4 µg/g (dry weight), respectively, but did not increase the concentration of selenium in muscle or kidneys. Burk et al.[84] reported similar concentrations of selenium in the necrotic liver of selenium-deficient rats.

The ratio of the concentration of selenium in liver to the concentration in kidneys increased from 0.2–1.0 to 2.8 to 3.7 as dietary supplementation with selenium increased from 100 to 500 to 2,500 ng/g.[314] The three successive fivefold increases in the concentration of dietary selenium (20 to 100, 100 to 500, and 500 to 2,500 ng/g) produced increases of 950%, 147%, and 30%, respectively, in the concentration of selenium in the liver. There was also an increase in the liver-to-kidney selenium ratio. Herigstad et al.[329] associated ratios greater than 1 in pigs with selenosis. The plateauing of selenium concentration with increased supplementation of the feed with selenium may be the effect of hemostasis or decreased food consumption. Since fecal excretion of selenium on the semipurified feed is less than 3% of the feed intake (compared with 30% in natural feeds), poor absorption should not have been a significant factor.

Maintenance of the concentration of selenium in rat muscle at the expense of the concentration of selenium in the liver was in contrast to observations in sheep, wherein the concentration of selenium in the liver (1 µg/g) was maintained at the expense of the concentration in muscle (0.5 µg/g, wet weight).[28,60,78,80] The relative ability of various species to spare selenium in the liver or muscle may affect the type or development of selenium-deficiency lesions (myopathy, hepatosis, and so on) in these species. There is an increase in the concentration of serum transaminase and dehydrogenases in selenium-deficient sheep and rats.[440,822]

Rats fed feeds containing 200–500 ng of selenium per gram were mated with selenium-depleted rats, but neither depleted males nor depleted females produced young by this method. Histologic sections of testicle and ovary from selenium-depleted rats did not contain normal numbers of viable sperm or oogonia.[724,818] Semen from selenium-depleted rats contained broken sperm, which lacked motility.[840,841]

Clinical infertility in rats is similar to observations of infertility in selenium-deficient sheep.[320] These conditions respond to selenium supplementation. They are associated with edema of the testicles, poor motility of the sperm, and (in lambs) myopathy.

Burk et al.[82] maintained rats for a month on a selenium-deficient feed with added α-tocopherol and then injected microgram amounts of ^{75}Se as selenite. The amount injected was equivalent to 100–500 ng of selenium per gram of body weight—the amount by which the experimental feed was deficient. Most of the injected selenium was retained by the rats for

Biologic Effects 91

6–8 weeks. Six weeks after inoculation, autoradiography and scintillation counting demonstrated that over 40% of the radioactive selenium was in the testicles and was concentrated in the midpiece of the sperm.

MONKEYS

Seven adult squirrel monkeys were fed a low-selenium semipurified feed with *Candida utilis* as the protein source and adequate vitamin E.[528] After 9 months the monkeys developed alopecia, loss of body weight, and listlessness. After one monkey died, three other monkeys were given 40 μg of selenium as selenite by injection at 2-week intervals. The three monkeys given selenium salts recovered. The three not given selenium became moribund or died. Lesions in these monkeys included hepatic necrosis, skeletal muscle degeneration, myocardial degeneration, and nephrosis. As in the rat, the tactile hairs of the monkeys were not affected by the loss of capillaries; they are supplied by cavernous blood sinuses rather than capillaries.

The homeostatic control of selenium retention and excretion appears to be quite close. Rats partly depleted of selenium and inoculated with 50 μg of selenium as selenite retained one-half of the injected selenium 6 weeks after inoculation.[82,345] The size of the inoculation was up to two times the amount of selenium "removed" from the diet during a preceding 30-day depletion period. Selenium-deficient ewes retained ^{75}Se in the erythrocytic portion of the blood for 150 days after injection of replacement amounts of selenium, 1 mg/50 kg of body weight.[816] Rats must be depleted through two or three generations of one or two litters to produce selenium-deficient young.[480] A fivefold increase of dietary selenium in semipurified feed, from 0.5 to 2.5 μg/g, increased the concentration of selenium in rat liver 30%.[314]

Animals can retain 25%–75% of dietary selenium consumed in natural feeds, but several factors influence this retention, including body stores of selenium, concentrations of selenium in the feed, level of intake, and the chemical form of selenium in the diet. Relationships between the level of dietary selenium and its concentration in animal tissues have been summarized by several authors.[8,28,60,87,327,377,406,408,416]

The concentration of selenium in tissues of poultry, rats, sheep, and swine maintained on selenium-deficient diets or those supplemented with 0.1–100 μg of selenium as selenite per gram of feed was not entirely dependent on the content of selenium in the feed.[28,60,80,87,314,329,345] Tissues from animals fed a selenium-deficient feed had more selenium in the kidneys than in muscle or the liver.[329] Sheep maintained the concentration of selenium in the liver;[80] rats tended to maintain the concentration in muscle.[314] Addition of physiologic amounts of selenium to the feed caused a proportional increase in the concentration of selenium in the liver (rat)[345] or muscle (sheep).[80] The ability of rats and other animals to effectively tolerate up to about 5 μg of selenium as selenite or selenate per gram of feed in natural foodstuffs, but not in semipurified feed,[313] may be due in part to combinations of selenium ions with components of the feed, or to formation of insoluble selenium compounds through reduction to selenide, or to precipitation of insoluble salts or complexes of metallic ions. These reactions are largely speculative; however, some physiologic reactions, experimental results, and observations suggest that the role of selenium in biology is broad and interrelated with other substances.

The Question of Selenium and Human Nutrition

Selenium is present in human blood and in all samples of human urine that have been analyzed by sensitive detection methods,[289,409,714] in human tissues,[656] and in a low-density serum lipoprotein from humans.[171] Concentrations in the tissues of California residents determined by X-ray emission spectrography[521 (pp. 119-125)] were generally similar to those in lambs and swine produced on commercial feeds.[186,646] Selenium levels in tissues of newborn infants in Russia[183] are about the same as those found in pigs and lambs on low-selenium diets in the United States.[535,646] The metabolic balance of selenium and other elements in New Zealand women was reported by Knight et al.[402] P. H. Weswig (personal communication) surveyed laboratory personnel and athletes in Oregon and found 150–300 ng of selenium per milliliter of blood. The concentration was independent of smoking habits, sex, exercise, and stage of training.

On the basis of the shape of the distribution curves of selenium in human tissues, Liebscher and Smith[444] concluded that selenium may be essential for man. There is no conclusive evidence that selenium deficiency is the specific cause of any human disease. Frost and others came to the same conclusion.[222,656]

Frost[222] and Schroeder et al.[656] estimate that the average human diet in the United States contains 1.8 mg of selenium per month (25 μg/kg of body weight per month). This compares with about 1 mg per month in a ewe whose lambs develop nutritional myopathy (20 μg/kg of body weight per month)[520,660] and with the 20 μg per month required for selenium-adequate rat diets (60 μg/kg of body weight per month).[480]

Clarification of the role of selenium is needed in several medical fields (discussed below).

Biologic Effects

KWASHIORKOR

Prompted by the initial report of Schwarz and Foltz,[666] Hopkins and Majaj[344] showed that administration of selenium to children in Jordan with kwashiorkor stimulated body growth and reticulocyte formation. In a discussion of this work, Burk reported findings in Guatemalan children with kwashiorkor who did not respond to the usual methods of nutritional supplementation.[344 (pp. 211–213)] The concentration of selenium in the blood of these children was 100 ng/g compared with 230 ng/g in well-nourished controls. When selenium salts were added to the previously unsuccessful kwashiorkor therapy, the health of the children improved, and the concentration of selenium in their blood increased to control levels. In children with low amounts of selenium, the uptake of selenium by erythrocytes was 21%, compared with 13% in children with normal amounts of selenium. Levine and Olson[437] reported concentrations of selenium in the blood of children with protein–calorie malnutrition that were similar to those of Burk et al.[81]

PERIODONTAL DISEASE

Periodontal disease is a major health problem in New England, a low-selenium area, and has been reported to be associated with selenium deficiency in sheep and cattle in New Zealand.[319,320] However, the concept that periodontal disease in sheep is a selenium-responsive disease was not confirmed in Scotland.[225]

SUDDEN INFANT DEATH SYNDROME

Although nutritional deficiency as a possible cause for the sudden infant death syndrome (crib or cot death) has not been widely considered, nothing excludes the possibility of a nutritional background for the event.[225,624] The studies of Money[495–497] suggest that sudden death in human infants may result from the combined deficiencies of vitamin E and selenium in cow's milk formulas. During the first month of life, breast-fed infants received more than 10 times as much vitamin E and more than two times as much selenium as infants fed cow's milk formulas.[486]

Rhead et al.[623] showed that the circulating levels of selenium and vitamin E in early infancy are low compared with those in adults. They concluded that, although vitamin E and selenium deficiency cannot be established as the primary cause of the sudden death syndrome, the possibility that nutritional deficits play a secondary role merits further investigation.

CARDIOVASCULAR DISEASE

Sudden death associated with selenium deficiency in newborn or rapidly growing lambs, calves, and pigs apparently results from weakening of the heart muscle. Selenium-deficient monkeys also have myocardial lesions. Hypoplasia of the vasculature of the skin has been demonstrated in selenium-deficient rats and monkeys. The primary degeneration of the sarcolemma area of the sarcomere and the secondary vascular degeneration in selenium-deficiency myopathy also suggest a cardiovascular function for selenium.

Although there is no evidence that selenium has a role in the maintenance of the cardiovascular system in human beings, Frost[222] compared maps of early heart mortality and cardiovascular-related deaths for different areas of the United States and demonstrated an inverse relationship between ambient selenium levels and the mortality pattern. Marjanen and Soni,[458] who hypothesized that manganese deficiency might underlie the very high cardiac and cancer mortality rates in Finland, have now adopted the view that selenium deficiency, which is prevalent all over Finland, may contribute to these unusually high death rates.

Lesions of selenium deficiency in rats and sheep have been associated with vascular abnormalities.[724] Demonstration of the role of selenium in maintenance of membranes may also suggest a function within the vascular system.

CANCER

Investigation of the direct relationship of selenium to human cancer has been limited to demographic studies and to comparisons of levels of selenium in the blood of patients with and without malignancies. Chu and Davidson[121] listed selenium compounds among potential antitumor agents. In addition, Shamberger and Rudolph[681] and Shamberger et al.[683] associated protection from cocarcinogenesis with antioxidants (vitamin E, selenium, etc.) and food preservation. Harr et al.[315] reported that concentration of dietary selenium delayed or prevented the induction of cancer by N-2-fluorenylacetamide (FAA). The effective concentration of dietary selenium in the torula feed in this experiment was the addition of 100–500 ng/g of feed.

Shamberger and Frost[680] published the first indication that human cancer mortality might bear an inverse relationship to selenium distri-

Biologic Effects

bution. Controlled animal studies conducted at the same time by Shamberger showing an inhibitory effect of selenium on carcinogenesis were not published until later.[679] Further investigation of the epidemiologic evidence for an inverse relationship between ambient selenium level and human cancer mortality included comparison of the levels of selenium in cow's milk with cancer mortality.[684] After regular use of selenium prophylaxis, including the addition of selenite to anthelmintic drenches, there was a rapid reduction in the incidence of ovine cancer in one part of New Zealand.[807]

On the average, the blood of cancer patients was reported to contain less selenium than the blood of other patients.[653,682] However, the blood of patients with some forms of cancer contained normal levels of selenium.[653,682,684] Patients with gastrointestinal cancer or metastases to gastrointestinal organs had significantly lower levels of selenium in the blood than normal patients. Mammary adenocarcinomas induced by FAA in selenium-depleted rats were more invasive than those induced in rats fed selenium-supplemented feeds.[315]

The ability of selenium to reduce methylene blue was reported by Schrauzer and Rhead[653] with the suggestion that this ability might provide a basis for testing for cancer, or susceptibility thereto. In studies of lipid therapy based on the types of lipid imbalance in cancer patients,[622] it was found that the most satisfactory and reproducible palliative effects of therapy were obtained by using synthetic lipids containing bivalent selenium, a serendipitous observation alluded to by Frost.[606]

REPRODUCTIVE SYSTEM

In the assessment of nutritional deficiencies as well as the effects of elements in excess, the reproductive system as a specific site of vulnerability is often ignored. The use of the radionuclide of selenium, ^{75}Se, has done much to elucidate the distribution of this element in the reproductive system, leading us to a greater awareness of possible specific roles for selenium in reproduction in the male, female, and developing progeny.

Distribution in Male Reproductive System

TRACER DOSES OF INORGANIC ^{75}Se

Until recent years there has been little reference to the distribution of selenium in reproductive organs, although Rosenfeld[635] reported that after repeated administration of tracer doses of ^{75}Se to rats, the testes contained

TABLE 5-2 ^{75}Se Uptake[a] in Various Tissues of CO-1 Male Mice following Subcutaneous Administration of Tracer Doses (1 μCi ^{75}Se; 0.03 μg Selenium)[b]

	Interval after Injection				
Tissue	1 Hr	4 Hr	1 Day	2 Days	7 Days
Kidney	11.19[c]	11.03	7.34	5.98	2.86
Liver	7.52[c]	7.28	5.55	5.23	3.61
Gastrointestinal tract	7.41[c]	4.97	2.79	1.86	0.63
Injected leg	3.32[c]	1.08	0.95	0.82	0.31
Blood	1.84	3.80[c]	1.68	1.27	0.74
Lungs	1.13	2.32[c]	1.50	1.35	0.76
Pancreas	0.95	1.16[c]	1.10	0.79	0.40
Spleen	0.82	1.34	1.76[c]	1.60	0.91
Heart	0.66	1.82[c]	0.88	0.83	0.51
Testis	0.56	0.67	1.37	1.58	1.97[c]
Noninjected leg	0.50	0.30	0.67[c]	0.56	0.32
Skeletal muscle	0.28	0.49[c]	0.37	0.31	0.19

[a] Uptake expressed as percentage of administered dose per gram of tissue. Mean values from five mice are shown.
[b] Derived from Gunn et al.[278]
[c] Highest concentration observed.

the highest concentration of selenium except the kidneys. Later, Gunn et al.[278] brought out that, after a single subcutaneous injection of a tracer dose of ^{75}Se to mice, the testes, which ranked low in ^{75}Se uptake at 1 hr, continued to cumulate this element, whereas all other tissues tested showed diminishing levels; by 7 days the male gonads ranked third (after the liver and kidneys) in ^{75}Se concentration (Table 5-2).

SPERMATOZOA AND SELENIUM DEFICIENCY

Brown and Burk[69] confirmed the cumulation pattern of ^{75}Se in the testes and epididymis of rats on torula yeast (low-selenium) diets and demonstrated by autoradiographs that ^{75}Se concentrated in the midpiece of sperm. They suggested that this may indicate a specific need for selenium in the mitochondria, which are found exclusively in the midpiece; subcellular fractionation showed that mitochondria of the testes contained more than twice as much ^{75}Se as those of the liver relative to the homogenate.

Although active spermatogenesis was observed in some of the seminiferous tubules of selenium-deficient rats born to females on a selenium-

Biologic Effects

deficient diet, the motility of spermatozoa from the cauda epididymis of these males was invariably very poor, most of the sperm showing breakage of fibrils in the axial filaments; these effects were not counteracted by the addition of vitamin E or other antioxidants.[841]

75Se IN ORGANIC FORM

Although no particular emphasis was placed on such observations, other investigators showed that there was a considerable accumulation of ^{75}Se when it was administered as the selenites selenomethionine or selenocystine in the testes and epididymis of mice[312] and sheep.[367] On the other hand, Anghileri and Marqués[19] drew attention to their observation that ^{75}Se, particularly when administered as selenocystine, continued to increase in concentrations in the testes of mice, while decreasing in other tissues. Along this same line, Patrick et al.[593] claimed that, in the fowl, administered ^{75}Se was bound to protein in spermatozoa, and the results of paper chromatographic studies suggested that it was present in part as selenocystine.

75Se WITH CARRIER SELENIUM

The failure of other investigators to observe high concentrations of ^{75}Se in the testes is attributable to the short-term nature of the distribution studies[345,652] and to a masking of the cumulation pattern in the testes by administration of the radioisotope with large amounts of carrier selenium.[467] Gunn et al.[278] showed, however, that upon administration of ^{75}Se with carrier selenium, a typical cumulative pattern of ^{75}Se occurred in the testes after a preliminary period of flushing out the initial high levels.

PERCENTAGE RETENTION OF 75Se IN THE TESTIS–EPIDIDYMIS COMPLEX

Brown and Burk[68,69] reported that the percentage of administered ^{75}Se retained by the testis–epididymis complex was far greater in rats reared on a selenium-deficient diet than had been reported by Gunn and Gould[275] for rats on regular diets (Figure 5-1). However, it is known that the retention of administered ^{75}Se in tissues and whole body is inversely related to the dietary level of selenium.[345,448] A far greater whole-body retention of ^{75}Se was reported by Brown and Burk[69] in rats fed a selenium-deficient diet than was observed by Blincoe[60] and by Gunn and Gould (unpublished data).

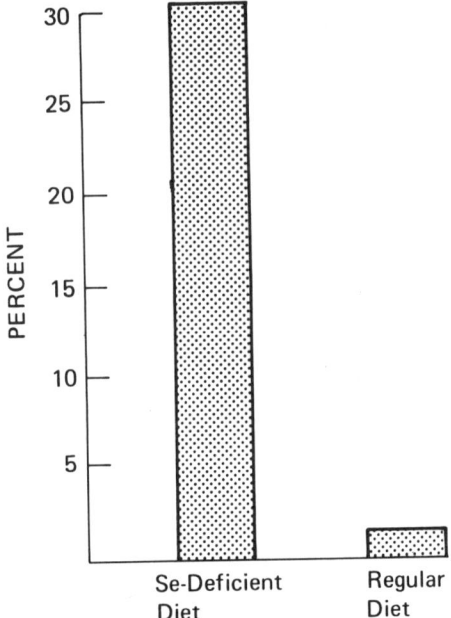

FIGURE 5-1 Percentage of administered ^{75}Se in the testis-epididymis complex of rats 3 weeks after administration of a single dose; comparison between rats on selenium-deficient diets (data of Brown and Burk[69]) and rats on regular diets (data of Gunn and Gould[275]).

NATURAL LEVELS OF SELENIUM IN HUMAN BEINGS

There is little information on natural selenium in animal or human testes. It is of interest, however, that in the data on human testes presented by Fuller et al.[227] and by Schroeder et al.[656] the testes ranked third among the tissues in selenium concentration, after the kidneys and liver. The selenium concentration in the testes of a 9-month-old child was about half of that reported for adult testis,[656] as would be expected if selenium were associated with spermatogenic elements in human beings.

INTERACTIONS WITH CADMIUM

Ganther and Baumann[238] showed that cadmium increased the retention of selenium in the body; the prolonged and increased retention pattern is also reflected in the testes.[276,279]

Kar and co-workers[386,387] and later others[64,279,282,463,464] showed that

Biologic Effects

the testicular damage induced by cadmium salts could be prevented by administration of selenium dioxide. Gunn and co-workers[276,279] investigated the mechanism of protection by selenium and found that selenium did not prevent cadmium from reaching the testes but instead caused a marked elevation in uptake of testicular cadmium. In view of the capacity of cadmium and selenium to augment concentrations of each other in the testes and the fact that at the same time selenium inactivates cadmium, it was postulated that some sort of binding existed between the two elements. Although the level of both elements initially increased, selenium later declined and cadmium continued to rise, which suggests that selenium has transported cadmium in an inactivated form away from the vulnerable site to some other locus within the testes where it is innocuous. Affinity-labeling studies with ^{109}Cd and ^{75}Se now confirm this hypothesis.[115,242] The plausible target of cadmium in cadmium-induced testicular injury was defined as a cadmium-binding protein with a molecular weight of 30,000 or possibly the crude nuclear fraction, or both. Following the administration of a protective dose of selenium, cadmium was diverted from these usual targets and, along with selenium, became attached to a protein of higher molecular weight.

Distribution in Female Reproductive System

The female gonads do not approach those of the male in capacity to concentrate administered ^{75}Se.[68,69] (See Figure 5-2.)

In analyses of human tissues, the ovaries ranked among the lowest in natural selenium content.[227] Generally speaking, there is a positive correlation between selenium intake and selenium content of tissues. Moxon and Poley[511] found that in hens fed seleniferous grain ration, the selenium content of the ovaries and oviduct was even higher than that of the liver and kidneys. Hen's eggs contain appreciable amounts of selenium, particularly in the yolk.[292,745]

PLACENTAL TRANSMISSION

More information is available concerning selenium in the female reproductive system during the pregnant state. Selenium is known to cross the placental barrier in several animal species.[312,335,368,476,635,815,834,837]

The finding by Muth et al.[526] that administration of selenium to the selenium-deficient pregnant ewe prevents the prenatal myopathy in developing lambs has been confirmed by many investigators.[521,641,769] Hadjimarkos et al.[294] also confirmed placental transmission of selenium in human beings. ^{75}Selenomethionine is now being used in pregnant

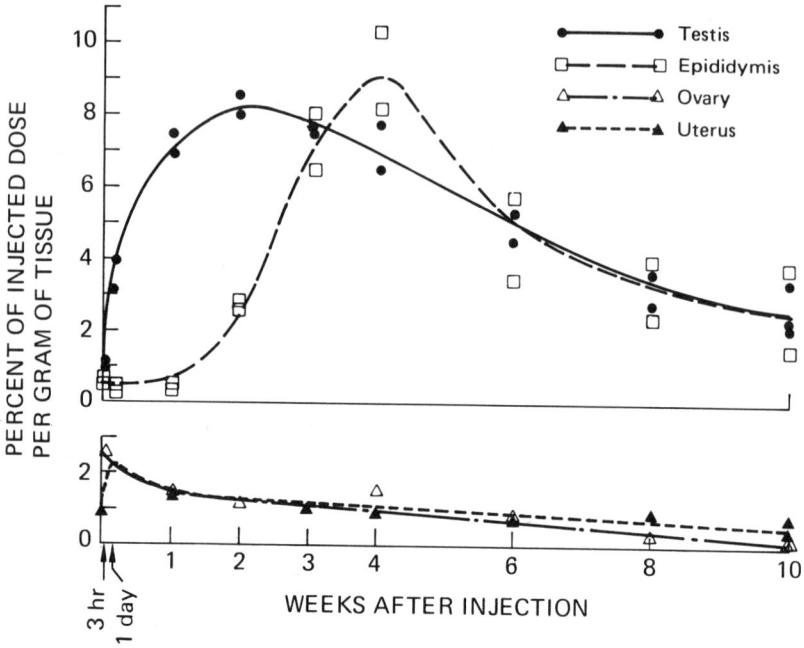

FIGURE 5-2 Comparison of retention of administered ^{75}Se in reproductive organs of male and female rats on torula yeast (selenium-deficient) diets. Each symbol represents one animal. From Brown and Burk.[69]

women to determine the efficiency of amino acid transport as an indicator of competence of the placental transport mechanism.[159,423] Although it is generally established that concentrations of selenium are less in the tissues of the fetus than in the mother, in the case of selenomethionine, selenium concentrations of fetal tissues approached those of the mother.[312,368] Ewes possess a carry-over effect from an adequate selenium intake during the first pregnancy to the second pregnancy when the dam is existing on a low-selenium ration that would otherwise produce lambs having white muscle disease.[817]

TRANSMISSION IN MILK

Selenium is also present in cow's milk in concentrations in direct proportion to the selenium intake.[12,244,267,293,597] Human milk contains selenium in concentrations twice as high as those in cow's milk.[288,486,656] In experiments with dogs, a single subtoxic subcutaneous injection of

Biologic Effects

radioactive selenium was sufficiently retained in the bitch to appear in the milk when lactation began after the second pregnancy.[476]

Recent studies in rats by Pařízek et al.[590] illustrate that mercuric salts induced a greater retention of selenium in the mothers, which resulted in reduced transfer of selenium to the fetus by both the placenta and the milk. These investigators emphasized the fact that surprisingly large amounts of mercury compounds have been used in seed dressings in agriculture;[49] as a result of increased concentrations of mercury in the environment connected with industrialization and agricultural development, the amount of selenium reaching the fetus might be diminished, contributing to a possible state of selenium deficiency.

Effect of Deficiency

The question of essentiality of selenium for specific reproductive processes is moot: deficiency of selenium interferes with the general health, and effects on reproduction may be secondary. Poor reproductive performance, expressed as low lambing or calving percentages, was frequently noted in sheep and cattle in areas where white muscle disease or "unthriftiness" was prevalent.[320,321] Estrus, ovulation, fertilization, and early embryonic development proceeded normally in affected flocks, but 3-4 weeks after conception, at about the time of implantation, embryonic mortality was high. Hartley et al.[321] first drew attention to the beneficial effects of selenium in correcting these deficiencies in reproduction. Buchanan-Smith et al.[78] reported that both selenium and vitamin E were required to obtain satisfactory reproductive performance in ewes fed a selenium-deficient, purified diet. Trinder et al.[763] cited the beneficial effects of selenium with vitamin E on the incidence of retained placentas in dairy cows on low-selenium diets. Not all workers praise the efficacy of selenium in correcting inadequacies in reproduction. Gardiner et al.[243] found that in areas of southwestern Australia, where selenium-responsive white muscle disease in sheep had been diagnosed, selenium had little or no effect on sheep fertility. Hill et al.[339] suggested that fertility, as measured by incidence of barrenness in 2-year-old ewes, was unaffected by selenium therapy, but that fecundity, as measured by twinning, was significantly increased. They suggested that the increased twinning was a secondary response to the increased weight of the selenium-treated animals, rather than a specific physiologic response to selenium. In some areas, infertility problems in sheep have been attributed to consumption of estrogenic pastures[148] rather than to selenium-deficient pastures.

Interference with reproduction is usually attributed to defects in the female. Buchanan-Smith et al.[761] detected no significant effects on reproductive organs of male lambs fed a purified diet very low in selenium for 140 days. When selenium-deficient diets were imposed over successive generations, rats showed adverse effects on reproduction. Animals grew and reproduced normally, but their offspring were almost hairless, grew more slowly, and were sterile.[480] The female progeny failed to reproduce when mated with normal males. The male progeny had immotile sperm with a morphologic defect of breakage of the axial filament.[841] Although these effects were alleviated by selenium, and not by vitamin E or other antioxidants, it cannot be stated categorically that this is a specific effect of selenium on male reproduction; the hairlessness and low body weights are indicative of generalized debilitation.

In the Japanese quail, deficiency of selenium in the diet resulted in reduced viability of newly hatched eggs, but the rate of egg production and fertility was not affected by the deficiency.[374]

Effect of Excess Selenium on Reproduction

DOMESTIC ANIMALS

Although there is no doubt that excess selenium has an adverse effect on reproduction, it has not been established whether these effects are specifically a selenium-toxicity effect or a secondary reaction to the accompanying emaciation.[641] According to Olson,[561] a reduction in reproductive performance is the most significant economic effect of chronic selenium poisoning of the so-called alkali disease type, and effects on reproduction may be quite severe without animals showing other typical lesions of selenosis. In the slightly more acute poisoning of the blind staggers type, both male and female gonads are affected. Rosenfeld and Beath[641] summarized the original observations of Draize and Beath[160] and Rosenfeld and Beath.[640] In the male the testicles are soft, flabby, and acutely congested; there is usually atrophy of seminiferous tubules. In adult females the ovaries are small, firm, and congested, usually with large numbers of atretic follicles and complete absence of mature follicles. In young females the ovaries are small, the corpus luteum is absent, and the number of follicles is greatly decreased. Changes in the gonads were also reported by Vesce[787] and Brusa and Oneto.[76] Rosenfeld and Beath[637] observed hypoplasia of reproductive organs in malformed lambs born to ewes that grazed on a seleniferous range. Wahlstrom and Olson[795] found that the feeding of 10-ppm selenite to young sows lowered the conception

Biologic Effects

rate, increased the number of services per conception, and increased the proportion of piglets that were small, dead, or weak at birth. Reports in 1955 from a district in Colombia described toxic corn and streams that had no animal life; small mammals using the streams for drinking water showed loss of hair and became sterile.[641]

LABORATORY ANIMALS

The effect of excess selenium on reproduction in laboratory animals depends on the duration of treatment and the dose of toxic substances.[641] Munsell *et al.*[518] reported that selenium-containing diets of rats had a detrimental effect on growth and reproduction in direct proportion to the selenium intake. Rats fed selenized wheat diets were usually infertile; matings in which one of the animals was normal were sometimes fertile, but affected females were unable to rear their young.[216,217] In studying the effect of concentrations of selenium in drinking water of 1.5 and 2.5 ppm, Rosenfeld and Beath[638] observed successful reproduction in two successive generations of male and female rats. The second generation of selenized rats, which received 2.5 ppm, had normal numbers of offspring, but only 50% of the young survived. An intake of 7.5-ppm selenium in water 5–8 days before parturition had no effect on the young before birth, but there was a decrease in the number of survivors with continued selenium intake.

Crossbreeding of selenized males and females with normal animals indicated that the fertility of the males was not affected, but the females failed to conceive or the few young born to them were unable to suck and appeared emaciated at birth. Schroeder and Mitchener[659] recently reported that mice given selenium in drinking water from weaning time reproduced normally until the third generation, which produced fewer and smaller litters, of which several were runts; also, deaths before weaning and failures to breed were excessive.

HUMAN BEINGS

Most reports on effects in human beings associated with exposure to excess selenium are concerned with pregnant women. The chief exception is a report from Japan that increasing numbers of female workers in the manufacture of selenium rectifiers had irregular menses or menostasis.[532]

Teratogenic Effects

CHICKS

The embryo of the chick is extremely sensitive to selenium toxicity. Hatchability of eggs is reduced by concentrations of selenium in feeds that are too low to produce symptoms of poisoning in other farm animals. Poor hatchability of eggs on farms has therefore proved to be an aid in locating potentially seleniferous areas where alkali disease in cattle, hogs, and horses may occur.[641] The eggs are fertile, but some produce grossly deformed embryos, characterized by missing eyes and beaks and distorted wings and feet.[105,211,219,274] Inherited abnormalities, such as the creeper mutation in hens, exaggerated the developmental malformations caused by selenium.[421] Deformed embryos were also produced by injection of selenite into the air cell of normal fertile eggs of both hens[211] and turkeys.[105] Kury et al.[411] suggested that in seleniferous areas the involvement of chick embryos could be more widespread than has been realized, not being confined to dead or grossly abnormal embryos. This conclusion is based on their findings of anemia (low red-blood-cell counts and hemoglobin values) in grossly normal as well as malformed embryos of chicks following injections of selenous acid into fertilized hen's eggs.

MAMMALS

The consumption of seleniferous diets interfered with the normal development of the embryo in many mammalian species, including rats,[216,638] pigs,[795] sheep,[637] and cattle.[155] In sheep, malformations of the eyes and of the joints of the extremities have been reported. The latter cause deformed legs and impaired locomotion.[641] These malformations were also observed in chicks. Holmberg and Ferm[343] did not observe teratogenic or embryotoxic effects in hamsters after intravenous administration of near-lethal doses of sodium selenite.

HUMAN BEINGS

Robertson[628] suggested that selenium may be a teratogen in man. Reports in the older literature of the people in Colombia eating toxic grains referred to malformed babies born to Indian women.[641] Robertson gathered information on the possible association between abnormal pregnancies and the exposure of women to selenite. Out of one possible pregnancy and four certain pregnancies among women exposed to selenite, only one pregnancy went to term, and the infant showed bilateral clubfoot. Of the

Biologic Effects

other pregnancies, two could have been terminated because of other clinical factors. Shamberger[678] cautions against using the inverse relationship between neonatal deaths and the level of selenium in some parts of the United States as a basis for a conclusion concerning the role of selenium in teratogenicity in human beings. Because of the many other factors in our environment that could influence the biologic availability of selenium, it appears that we would be unjustified in concluding, solely on the basis of this evidence, that selenium has no bearing on teratogenicity in human beings. Rosenfeld and Beath[641] emphasized that studies of mammalian malformations in relation to the age of the embryo or fetus and its susceptibility to selenium would be of great value to basic as well as applied research.

Effect of Supplementation

Although some investigators have reported that supplements of selenium, either alone[320,321] or in combination with vitamin E,[78,763] corrected reproductive deficiencies in animals with white muscle disease, others[148,243,339] have reported that selenium supplements did not improve reproductive performance in some areas where selenium-responsive white muscle disease was prevalent. On the other side of the balance, there have been reports of adverse effects from supplements of selenium. Grant[266] reported some interference with conception in ewes grazing New Zealand pastures sprayed with 50 g of selenium as selenite per acre.

VASCULAR SYSTEM

Selenium produces widespread toxic effects, with many organs of diverse species being affected. There is no agreement on a mode of action that could explain this multitude of toxic reactions. In reviewing what is known of the manifestations of selenium excess and deficiency, one is struck by the vascular characteristic of selenium disorders.[276] The question of a primary vascular involvement in selenium imbalance has been largely ignored. Indeed, consideration of the vascular endothelium as a dynamic structure, with its own selective reactions to injury, is a relatively new concept and opens up a whole new field of investigation.[277,284]

Selenium Deficiency

In certain selenium-responsive deficiency disorders, such as exudative diathesis in chicks, the primary effect is considered to be on capillary permeability.[50,140] However, one cannot exclude the possibility that even necrotic liver degeneration in rats,[663] cardiac and peripheral muscle degeneration, liver and kidney necrosis, and pancreatic dystrophy in mice,[153] and the various myopathies in Herbivora[641] could have the vascular endothelium as their primary involvement.

Recent experimental evidence bears out this suggestion. In studies of ultrastructural changes in the hearts of piglets from sows deprived of vitamin E and selenium, Sweeny and Brown[739] noted that lesions appeared first in connective tissue and in capillaries, preceding any apparent structural changes in muscle cells proper. Sprinker et al.,[724] in studies with rats on selenium-deficient diets, noted vascular hypoplasia and endothelial thickening and degeneration in tissues with a marginal vascular supply and oxygen dependence, such as skeletal and cardiac muscle, testes, and retina. Since there were widespread vascular lesions and secondary degenerative lesions in several vascular-dependent tissues, the authors concluded that selenium deficiency caused primary damage to the vasculature and also had an influence on the maintenance of membranes.

Supplee[734] described a defect in the flight feathers of poults fed a diet deficient in vitamin E and selenium. This defect resulted in discoloration and atrophy, which he attributed to degenerative changes following hemorrhage in the pulp of the immature feather. The vascular system is further implicated in a description of antiinflammatory properties of selenium compounds,[627] which subsequently found application in the use of a selenium–tocopherol treatment for chronic lameness in dogs.[127]

Money[495] suggests that the young of about 40 mammalian and bird species cannot tolerate vitamin E and selenium deficiency. Despite the disparateness of their various fatal diseases, all the diseases are characterized by effusion, hemorrhage, and necrosis of cells.

Selenium in Excess

ACUTE SELENIUM POISONING

Vascular manifestations are most apparent in selenium poisoning. The following statements are extracted from the original observations of Draize and Beath[160] and Rosenfeld and Beath[640] on acute selenium poisoning in farm animals. Petechial hemorrhages appeared in the endocardium of the heart. The lungs showed acute congestion and diffuse hemorrhages. The omasum showed congestion, hemorrhages, and desquamation of epithelium of the mucous folds. The intestines were hemorrhagic and showed enteritis and occasionally colitis and proctitis. The liver showed passive congestion, hemorrhages, and parenchymatous degeneration accompanied by focal necrosis. The kidneys showed parenchymatous degeneration and hemorrhages accompanied by nephritis. The spleen was acutely congested. Microscopic study showed acute congestion in the endocardium, focal necrosis, and hemorrhages; the pericardium exhibited petechial hemorrhages. The lungs revealed hemorrhages in the alveoli and occasionally in the interstitial tissue. The mucosa and submucosa of the stomach and intestines manifested edema, hemorrhages, and necrosis. The capillaries of the liver lobules were dilated and congested. The kidneys showed parenchymatous degeneration. The pancreas, gall bladder, spleen, and lymph nodes showed congestion and hemorrhage.

CHRONIC SELENIUM POISONING OF THE SO-CALLED ALKALI DISEASE TYPE

Lesions of so-called alkali disease represent chronic progressive degeneration.[160,218,503,640] Petechial hemorrhages were seen on the epicardium. The lungs showed focal fibrosis, early congestion, and some edema.

CHRONIC SELENOSIS BY INORGANIC SELENIUM

A few subendocardial hemorrhages have been noted in equines. In the kidneys the cortex showed hemorrhages, and the medulla was congested. Microscopic observation of the liver showed focal necrosis, fatty infiltration, congestion, edema, cellular changes with cloudy swelling, and complete loss of cellular structure.[491]

ACUTE SELENIUM POISONING IN LABORATORY ANIMALS

In an early report, Jones[381] described the right auricle of the heart as distended and full of clotted blood; the splanchnic vessels were enormously dilated. Toxic edema of lungs and parenchymatous organs was noted in rats exposed to high concentrations of selenic anhydride.[198] Rats, guinea pigs, and rabbits exposed to selenium fumes gave no evidence of injury except on the lungs, which were hemorrhagic.[297] Investigating the effects of various chemical elements on permeability as tested by local intracutaneous injection in guinea pigs with circulating Evans blue, Steele and Wilhelm[728] demonstrated that selenite had striking activity in increasing vascular permeability.

SUBACUTE AND CHRONIC SELENOSIS IN LABORATORY ANIMALS

The first description dealing with pathologic changes related to chronic selenium poisoning was provided by Duhamel.[167] The lungs showed hemorrhagic exudate in the alveoli, dilated capillaries, and bronchial exudate. Necrosis, hemorrhage, and fibrosis were the essential stages in the histogenesis of hepatic cirrhosis. Telangiectasia with focal necrosis was frequently present in some stages of liver damage.[109] In the kidneys, there was usually mild tubular degeneration with acute glomerular injury. Pathologic changes have also been described by Franke,[208] Munsell *et al.*,[518] and Smith *et al.*[713] In a study of chronic selenium intoxication in dogs, Moxon and Rhian[513] noted small local hemorrhages, severe ascites, liver and splenic damage, emaciation, apathy, and progressive anemia.

In early cases of subacute selenosis, the most prominent feature is dilation of veins in the visceral region. The vena cava and right auricle are always enlarged. The lungs and liver are congested. The stomach and intestinal tract show hemorrhages. The bladder is distended and filled with colored urine. In the truly chronic poisoning, the outstanding pathology occurs in the liver. Heart and spleen are enlarged. Lymph nodes are enlarged and congested. Ascites and edema are common. In guinea pigs, poor subperiosteal ossification and hyperemia at the boundaries of primitive cartilage and hyperemia of the parathyroid gland are present.[33]

The vascular effect is apparent even in nonmammalian species. In reporting on goldfish poisoned with selenium, Ellis *et al.*[180] described marked edema of many tissues, particularly the submucosa of the stomach and around the blood vessels in the kidneys and liver.

In studies of embryonic malformations in eggs laid by selenium-fed hens, Gruenwald[274] concluded that the defects resulted from regression of previously well-formed parts rather than from an interruption of the normal developmental process. Tissue necrosis in certain areas of brain, spinal cord, eyes, and limb buds was a constant feature. His observations of hemorrhage, coupled with the knowledge that brain tissue has a marginal vascular supply,[724] suggest the possibility of interference of selenium with the vascular system even at this embryonic stage.

SELENIUM TOXICITY IN HUMAN BEINGS

In a survey of rural populations living in seleniferous areas, subcutaneous edema (probably of cariorenal origin) was seen in five cases.[641] With exposure to selenium dioxide powder, the cases of inflammation of the nail beds are especially painful.[112] A case of acute lethal poisoning in a 3-year-old boy is cited by Carter,[107] who noted dilation and increased permeability of the peripheral vasculature, presumably at the capillary level, and congestion and edema of the lungs and gastrointestinal tract. Carter pointed out that the effect of selenium intoxication on the peripheral vasculature is similar to the effect of arsenic.

The Vascular System as a Vulnerable Site for Toxicity

In various metabolic studies, vascular tissue has seemingly been ignored. Yet, recent studies have shown that the vascular system is more than a simple conduit for blood; it is a dynamic system playing a vital role in selective permeation of nutrients to tissues; in turn, the endothelium may have selective reactions to injury and may well be the primary site of damage from various forms of noxious stimuli.[14,277,284,845] Certain nephrotoxic snake venoms, mercuric and uranium salts, and chromates have been shown to exert their renal damage by means of selective damage to the glomerular vasculature.[14,354] Arsenic is known to interfere with capillary permeability, an effect that can be prevented by sulfhydryl compounds;[143] it is of interest too that arsenic is an antagonist to many of the toxic effects of selenium.[17] Cadmium, which recently has been suspected as a culprit in hypertensive vascular diseases,[654] has also been implicated in some other vascular reactions that involve selenium. In the male, the overwhelming necrosis that subcutaneously administered cadmium salts provoke specifically in the testes[275,587] has been shown to be due to selective damage to their vasculature and to the internal pampiniform plexus complex and branches of the spermatic artery.[275,284] Although relatively large amounts of zinc,[285,587] sulfhydryl compounds,[280] and estrogens[281] can protect against this specific vascular damage, the most potent known protector is selenium.[278,279,282,283,387,463] Cadmium also evokes an acute hemorrhagic necrosis in the placentas of rats[588] and mice[118] near term, and some have claimed that the ovaries of prepubertal rats also undergo an acute, though transient, hemorrhagic response that can be blocked by zinc and selenium.[387]

The review of essentiality and toxicity of selenium points to the vascular system as one of the tissues with which this element is involved. Further research is definitely needed to assess the role that selenium may play as a nutrient in the regulation of and as a toxicant in the destruction of the barrier that governs the health of all tissues.

SELENOSIS

Although selenium was reported to be toxic to animals as early as 1842, it was not until the early 1930's that its role as a poison received much attention. At this time, the U.S. Department of Agriculture and the Wyoming and South Dakota agricultural experiment stations found this element to occur at toxic levels in plants growing on certain soils of the Great Plains and Rocky Mountain area and to cause the so-called alkali disease and certain heretofore unexplained losses of livestock. Several excellent reviews of this early work are presented elsewhere.[18,316,503,636,756] Since then, there has been increased concern over the toxicity of selenium.

Types of Selenium Poisoning

Following the discovery of the relation between selenium and livestock losses, it was soon recognized that selenium poisoning has more than one form, and it is important that this be kept in mind during any discussion of the toxicity of the element. Rosenfeld and Beath[641 (pp. 141-162)] suggested three types of field selenium poisoning in livestock: acute, chronic of the blind staggers type, and chronic of the alkali disease type. These are briefly discussed below.

In the field, acute poisoning occurs when grazing animals eat sufficient quantities of the selenium accumulator plants to cause sudden death or signs of severe distress, such as labored breathing, abnormal movement and posture, prostration, and diarrhea. Since animals usually avoid these plants, this type of poisoning is rare. In periods of pasture shortage, however, accumulators are sometimes about all that is available to eat, and occasional losses of large numbers of grazing sheep and cattle from the acute form of poisoning have been reported.[35] The acute poisoning has also been produced experimentally or accidentally by the administration of selenium compounds to farm animals.[102,103,230,329,420,492,500,539,575,636,688]

Chronic selenium poisoning of the blind staggers type has been reported to result from the ingestion of accumulator plants in limited amounts over a period of weeks or months.[636] Affected animals wander, stumble, have impaired vision, and show some signs of respiratory failure. This type of poisoning has been produced experimentally by the administration of water extracts of accumulator plants, but not by the administration of pure selenium compounds. Since the plants from which the extracts were prepared probably also contained some toxic alkaloids, it has been pro-

Biologic Effects

posed that these rather than the selenium produced the poisoning.[454] This matter needs further study.

Chronic selenium poisoning of the alkali disease type has been discussed in detail by Moxon[503] and by Rosenfeld and Beath.[636] It results from the ingestion of feeds containing toxic amounts of selenium over weeks or months. Toxic amounts are different for different animals, ranging from about 5 ppm to about 40 ppm. The most obvious signs of the poisoning are: in cattle and horses, lameness, hoof malformation, loss of long hair from mane or tail, and emaciation; in swine, lameness, hoof malformation, loss of body hair, and emaciation; in poultry, decreased hatchability of the eggs due to teratogeny. Sheep have not been observed to show hoof or wool lesions, but their reproduction is adversely affected,[155] and this has also been observed in cattle,[493] swine,[795] and rats.[216] Selenium as a cause of alkali disease has also been questioned,[454] but experiments have demonstrated that signs of the syndrome are caused both by grains or grasses of high selenium content and by inorganic salts of the element.[211,216,490,503,566,625,794]

As more studies on the toxicity of selenium are reported, it becomes obvious that the effects of the element gradually increase in severity as selenium intake increases, so that it is difficult to differentiate clearly between the different forms of the poisoning. Thus, the lesions or signs of the toxicosis reported in the literature may vary widely, even within a single species, and no attempt will be made here to review them in detail. Several authors in addition to those cited here have reported on this matter, and the reader is referred to the original reports for details.[85,103,200,329,651,688,784]

Factors Affecting Selenium Toxicity

The literature is replete with illustrations of conditions or factors that alter the effects of selenium on animals. These in turn have affected the results of research designed to establish toxic levels, complicating, in particular, efforts to establish a "no-effect" level for the element. Some of these conditions and factors are discussed below.

The route or method of administration can be expected to cause real differences in measurements of toxicity. Methods include intravenous, intraperitoneal, or subcutaneous injection; administration in the food or water; application to the skin; and subjection to vapors.

The rate of intake or administration also has a great effect and makes the difference between no effect, chronic effects, or acute effects. The effects of continuous selenium intakes at apparently nontoxic levels over

TABLE 5-3 Acute Toxicity of Some Selenium Compounds

Compound	Experimental Animal	Mode of Administration	Toxicity[a]	Reference
Sodium selenite	Rat	Intraperitoneal injection	MLD[b] 3.25–3.5 mg Se/kg body wt	210
	Rat	Intravenous injection	MLD[c] 3 mg Se/kg body wt	714
	Rabbit	Intravenous injection	MLD[c] 1.5 mg Se/kg body wt	714
	Rat	Injection	MLD 3–5.7 mg Se/kg body wt	513
	Rabbit	Injection	MLD 0.9–1.5 mg Se/kg body wt	513
	Dog	Injection	MLD 2 mg Se/kg body wt	513
Sodium selenate	Rat	Intraperitoneal injection	MLD[b] 5.5–5.75 mg Se/kg body wt	210
	Rat	Intravenous injection	MLD[c] 3 mg Se/kg body wt	714
	Rabbit	Intravenous injection	MLD[c] 2–2.5 mg Se/kg body wt	714
Selenium oxychloride	Rabbit	Application to skin	83 mg of compound caused death in 5 hr; 4 mg caused death in 24 hr	164
Hydrogen selenide	Rat	In air	All animals exposed to 0.02 mg/liter of air for 60 min died within 25 days	165
DL-selenocystine	Rat	Intraperitoneal injection	MLD[a,b] 4 mg Se/kg body wt	507
DL-selenomethionine	Rat	Intraperitoneal injection	MLD[a] 4.25 mg Se/kg body wt	401
Diselenodipropionic acid	Rat	Intraperitoneal injection	LD$_{50}$ 25–30 mg Se/kg body wt	508
Dimethyl selenide	Rat	Intraperitoneal injection	LD$_{50}$ 1,600 mg/kg body wt	475
Trimethylselenonium chloride	Rat	Intraperitoneal injection	LD$_{50}$ 49.4 mg Se/kg body wt	553

[a] MLD, minimum lethal dose; LD$_{50}$, dose causing death in one-half of test animals.
[b] Smallest amount that would kill 75% of the rats in less than 2 days.
[c] Smallest amount that would kill 40%–50% of the animals.

Biologic Effects

a period of years have not been well documented, and this is a matter of importance in the case of man.

Different species are affected by selenium in different ways, and some species are more resistant than others. Thus, man may not respond to toxic doses as livestock or experimental animals do, and extrapolation of results from animals to human beings must be done very cautiously.

Young animals seem more susceptible to the poisoning than do older ones, and embryos—especially the embryos in eggs—seem even more susceptible. Different chemical forms of selenium have greatly different toxicities (Table 5-3).

Criteria Used in Measuring Toxicity

The literature reveals a number of criteria that have been used in determining toxicities of the various forms of selenium. These include death of the animal, signs suggesting severe distress or pain, impaired breathing or vision, impaired movement, gross lesions (both external and internal), biochemical changes, cariogenesis, and microscopic pathologic changes. Because of these widely differing criteria, it has been difficult to determine the amount of selenium that constitutes a toxic dose and the level at which intake or exposure becomes harmful.

Acute Toxicity of Selenium

Table 5-3 summarizes some of the data reported in the literature on the acute toxicity of various selenium compounds. The data illustrate the wide differences in toxicity of the various chemical forms of the element, some species differences, and various ways of expressing toxicity. Although the possibility of acute selenium poisoning exists in some areas for range animals and in certain industries for man, the problem of acute toxicity seems less important than that of chronic toxicity.

Chronic Toxicity of Selenium

In 1937, Moxon[503] stated that, in general, it could be said that selenium poisoning of the alkali disease type results when animals consume feeds containing 5–40 ppm of the element over a period of time. There are numerous reports that diets containing 5 ppm or more do indeed cause chronic toxicity, and in seleniferous areas this has been accepted as the dividing line between toxic and nontoxic feeds.

However, the suggestion that something less than 5 ppm of the element in the diet can be toxic is often seen in the literature. It is also suggested that selenium in water may be more toxic than that in feed. The discussion that follows relates primarily to work that attempts to establish a "no-effect" dose level for the element and thus arrive at some conclusion as to what levels in feed or water can be expected to harm man, wildlife, or livestock.

In 1967, Tinsley et al.[754] concluded that, so far as longevity is concerned, a daily dose of 0.5 mg of selenium as selenite or selenate per kilogram of body weight per day seemed to be the threshold dose in rats on a casein-Cerelose diet (for a 200-g rat eating 10 g of feed per day, this would be the equivalent of 10 ppm). On the other hand, a calculated maximum body weight was reported to be decreased by as little as 0.5 ppm of selenium. In addition, Harr et al.[313] reported that when additions of 0.5-2 ppm of selenium were made to the diets, proliferation of the hepatic parenchyma was more prevalent than in control animals on diets with no added selenium and that selenium added to a commercial diet produced less toxicity than selenium added to a casein-Cerelose diet.

A complementary report gives detailed data.[51] Here again, the weight effects were noted. However, a careful study of the data on chronic liver and bile duct hyperplasia (see diet 1, p. 55; diet 2, p. 57; and diet 21, p. 94) shows that this lesion was even more prevalent in a commercial diet without added selenium than in a casein-Cerelose diet with 0.5 ppm of added selenium. This may mean that the hyperplasia does not indicate a toxic effect of the element. In a later report, Harr and Muth[316] state, with reference to the studies of the semipurified diet, that the minimum toxic level for liver lesions was 0.25 ppm. With reference to longevity and lesions in heart, kidneys, and spleen, they concluded that the minimum toxic level was 0.75 ppm. They state, however, that rats fed 0.5 ppm of selenium in the diet grew as well as the controls. They concluded that the estimated dietary threshold for physiologic-pathologic effects is 0.4 ppm and for pathologic-clinical effects, 3 ppm. Neither growth nor longevity was adversely affected by as much as 2.5 ppm of added selenium in a torula yeast diet to which the carcinogen fluorenylacetamide had been added. The physiologic significance of some of the observations of this group is difficult to evaluate.

Pletnikova[611] has recommended a maximum concentration of 0.001 mg of selenium as selenite or selenate per liter of water for Russian drinking water. She reports 0.01 mg/liter as the threshold for detection by odor. She also reports decreased liver function and effects on the activities of some enzymes along with increased blood glutathione in rats receiving 0.5 μg of selenium per kilogram of body weight per day (about 0.01 mg/liter) for a period of 6 months. These effects were not obtained at a level of one-tenth of this amount. Unfortunately, she does not describe the diet or state its selenium content. Quite likely, the selenium intake from it was considerably greater than that from the water containing 0.01 mg/liter. Further, bromsulphthalein (BSP) clearance was used for the liver-function test. With this, BSP is excreted into the bile conjugated with reduced glutathione (GSH). If selenium catalyzes GSH oxidation, the GSH pool available to react with the dye would be depleted; hence, the effect may not indicate a toxicity. The physiologic significance of the observations made in this study is not clear.

Palmer and Olson[586] studied the toxicity of selenite and selenate to rats on corn- or rye-based diets. They administered selenite or selenate in water at the rate of 2 or 3 mg/liter for a period of 6 weeks. Each form produced a small reduction in rate of gain without mortality. Earlier, Schroeder and Mitchener[658] reported severe toxicity at the 2 mg/liter level for selenite selenium but not for selenate selenium.

Halverson et al.[303] fed postweanling rats for 6 weeks on wheat diets containing 1.6, 3.2, 4.8, 6.4, 8.0, 9.6, or 11.2 ppm of naturally occurring or selenite selenium. Growth was not affected below the 4.8-ppm level of selenite or the 6.4-ppm level of selenium from grain. At 6.4 ppm of selenium or above, restriction of feed intake, increased mortality, increased spleen weight and size, increased pancreas size, reduced liver : body weight ratios, and reduced blood hemoglobin were noted. These effects were not observed in rats on diets containing lesser amounts of selenium.

Thapar et al.[747] found that 8 ppm of selenium added as selenite to either a practical corn–soy diet or a Cerelose–soybean protein diet reduced egg production, weight and hatchability of eggs, body weight, survival rate, and growth of progeny of laying hens fed the diet from 1 day of age for as long as 105 weeks. But no detrimental effects were observed when selenium was added at the rate of 2 ppm, and it is possible that this addition improved the livability of hens on the Cerelose–soybean protein diet. Similar findings were later reported by Arnold et al.[22] Much earlier, Poley et al.[612] reported that 2 ppm of selenium from grain improves the growth of chicks on a practical-type diet.

Witting and Horwitt[832] reported that growth curves had shown that the selenium requirement of the tocopherol-deficient rat has a very narrow optimal range. The best growth rate was obtained on the addition of 0.1 ppm of selenium as selenite. At 0.3 ppm of selenium, the growth was better than at 0.03 ppm, but not as good as at 0.1 ppm. With the diet severely deficient in vitamin E, selenium toxicity was noted at what these authors considered an unusually low level of the element: 0.25 ppm in the basal diet plus 1 ppm as selenite.

Obviously, the chronic toxicity of selenium will depend on the criteria used to determine the "no-effect" dose level. For the normal diet, 4–5 ppm will usually inhibit growth, and this may be the best indicator of toxicity. In a diet deficient in vitamin E, 1 ppm may be toxic. During the development of teeth, 1-2 ppm may be toxic if subsequent cariogenesis is used to measure toxicity. Histopathologic observations may suggest that less than 1 ppm can be toxic. However, the physiologic significance of the observations may not be clear, and the same may be said for biochemical parameters indicating that even lower levels can be toxic. In many areas, livestock are regularly fed diets containing over 0.5 ppm of the element, and there has been nothing to suggest that they fare less well than animals on diets of lower selenium content.

Treatment or Prevention of Selenosis

Rosenfeld and Beath[640,641] state that no treatment is known for counteracting the toxic effects of large amounts of selenium. Therapeutic measures, therefore, are entirely symptomatic. It should be pointed out that British Anti-Lewisite (BAL) is contraindicated[66] and that the use of ethylenediaminetetraacetic acid (EDTA) is essentially without effect.[706]

Strychnine sulfate and prostigmine have been used with some success in the treatment of the blind staggers syndrome,[34,636] but in view of the uncertainty over the cause of this type of poisoning, there is doubt that these are treatments for selenosis.

Olson[561] has reviewed the various methods suggested for the prevention of chronic selenium poisoning in livestock. At present the only practical method is to avoid allowing the animals an excessive intake of selenium. Affected animals that are removed from seleniferous feeds may recover without apparent aftereffects. The rate and extent of recovery depend largely on the severity of the damage suffered by the animals before removal from the toxic feeds.

Selenium Poisoning in Man

Some reports of selenium poisoning in man have been concerned with excessive dietary intake of the element in foods and have suggested that signs of poisoning include chronic dermatitis; loss of hair; and loss, discoloration, or brittleness of fingernails.[424,425,636,674] The evidence presented, however, does not establish that selenium was the cause of the signs observed in the studies.

Other reports have dealt with selenium poisoning in industrial workers,[305] but no death among these workers has been attributed to such poisoning. Workers exposed to fine dust of elemental selenium, which is

Biologic Effects

very insoluble in water, have suffered catarrh, nosebleed, and loss of sense of smell in instances where the dust collected in the upper nasal passages. A dermatitis was observed on the hands of a few of those handling the element.[16] Molten selenium has reportedly caused burns without any toxic reaction,[615] and in one instance where rectifier plates made of the element were melted down, the red fumes caused intense eye, nose, and throat irritation, with some dizziness and headache.[124]

Hydrogen selenide has also been implicated in the poisoning of man in industrial situations.[257,740] Irritation of the mucous membranes of the respiratory tract, nausea, and dizziness are among the reported signs of poisoning. Hydrogen selenide has not caused a death in man; one reason may be that it is never used in quantity, and another may be that it is readily oxidized to give elemental selenium.[257] At concentrations as low as 0.001 mg/liter of air, it causes olfactory fatigue, and its offensive odor cannot be relied upon to warn of its high concentration. As little as 0.005 mg/liter of air has been reported intolerable to man,[165,166] and the threshold limit given by the American Conference of Governmental Industrial Hygienists is 0.05 part of the gas per million parts of air (by volume).[15]

Selenium dioxide forms selenious acid when in contact with water and is the main problem in industries using selenium. The sudden inhalation of large quantities of selenium dioxide powder may produce pulmonary edema because of the local irritant effect on the alveoli of the lungs, and persons exposed to it may experience mild epigastric distress after meals.[257] The compound has caused dermatitis[615] and burns when in contact with the eyes.[484] If allowed to penetrate beneath the fingernails, it causes an especially painful inflammatory reaction.[258] Glover[257] states that the symptoms of overexposure to selenium in man—pain in the nail beds, metallic taste, and garlic odor of the breath—make workers treat selenium compounds with respect and may lessen the chances for accidental poisoning.

Selenium oxychloride, an almost colorless liquid at room temperature, is a severe vesicant capable of producing a third-degree burn, penetrating the skin, and appearing in the blood and liver.[164] Acute sore throats in laboratory workers exposed to dogs exhaling dimethylselenide after selenium injections have been reported,[502] but additional data on the toxicity of organic selenium compounds to man are not available. Studies of a community in the neighborhood of a selenium-refining plant in Japan indicated that a yellow color of the facial skin, anemia, and hypertension in some subjects might have been signs of selenium toxicity.[736] And finally, Frost[223] has suggested that the British beer-poisoning epidemic of 1900 may well have involved selenium.

Until 1966, the literature had not carried any report of the death of a human being as a result of accidental exposure to selenium. In that year, Carter[107] reported the fatal poisoning of a 3-year-old boy who ingested selenious acid in a proprietary preparation of a gun-bluing compound. The pathologic findings suggested a toxicity of the vasculature. Carter contended that general unawareness of the possibility and mode of presentation of acute selenium poisoning renders its diagnosis unlikely and tends to perpetuate the appearance of rarity. Selenium disulfide, a commercial preparation used in the treatment of dandruff, has been suspected of causing a nonfatal toxicity in a woman using it. Red lipstick containing selenium has been suggested as the cause of a syndrome in a patient using it.[77] The syndrome expressed itself in nervousness, metallic taste, vomiting, mental depression, and pharyngitis.

Dental Caries

Hadjimarkos and his co-workers have published many articles on the relationship of selenium to dental caries,[290] beginning in 1952 with a report on a study in which it was concluded that a direct relationship existed between dental caries in man and urinary excretion of the element.[296] In reviewing this subject in 1968, Hadjimarkos[287] states that epidemiologic studies among children and experimentation indicate that selenium is indeed capable of increasing the susceptibility of teeth to decay. He concludes that the ingestion of the element during the development of teeth probably results in its incorporation into the tooth structure and a subsequent increase in the incidence of caries.

Tank and Storvick[742] reported that children whose urine contained more than 0.1 ppm of selenium had an increased incidence of caries. Cadell and Cousins[98] observed no correlation between selenium content in urine and the incidence of caries in children, but Hadjimarkos[291] pointed out that in no case was the urinary selenium excretion in the children studied at a high enough level to suggest that increased caries could be expected. Ludwig and Biddy[451] found that although certain towns in low-selenium areas had a lower incidence of dental caries than did some in high-selenium areas, one of the towns in a high-selenium area had the lowest incidence of caries. These investigators also pointed out that the incidence of caries in all the high-selenium areas was lower than that reported for towns in New England, a low-selenium area. Comments by Schwarz[664] emphasize that surveys of the type mentioned above, as well as the attempt to draw any meaningful conclusions from the early reports that families in seleniferous areas had bad teeth,[711] are deceptive, because they do not take into account the many factors that might influence the incidence of dental caries, such as economic status or level of education.

A variety of results in addition to those reported by Hadjimarkos and his co-workers have been obtained in experimental animals. Claycomb et al.,[122] Muhler and Shafer,[517] Navia et al.,[537] and English[181] found the administration of selenium to be essentially without effect on the incidence of dental caries. Mühlemann and König[516] reported a reduction in the incidence of caries in rats fed selenium, attributing this to a reduction in consumption of the cariogenic diet. Administering selenium to young rats after weaning and to their dams during pregnancy and lactation increased the incidence of caries in the young.[90] In agreement with this and with some of the work of Hadjimarkos,[287] Bowen[65] reported that, after the administration of selenium to monkeys in their water during tooth development, chalky yellow enamel developed. When it was administered posteruptively, the selenium seemed to be cariostatic. Further, Shearer[685] found that the uptake of selenium from selenomethionine was much greater during development of teeth than it was after the teeth were fully developed. Unfortunately, the work with experimental animals has dealt with selenium added to water or feed at levels that are quite toxic and well above what the general population would be expected to consume. It is dangerous and unrealistic to extrapolate the findings to man.

This matter is worthy of additional study, but at present there seems no reason to suspect that selenium is important to cariogenesis in man.

CARCINOGENICITY AND ANTICARCINOGENICITY

Since 1940, five groups of scientists have studied the carcinogenic and anticarcinogenic potential of selenium, selenium salts, mixed selinides, and seleniferous grains. The first group fed a stock colony of rats seleniferous grain and a mixed selenide insecticide; the second group used natural feed with added selenium salts; the third and fourth groups used semipurified rations with added selenites and selenates; and the fifth group used natural feed with selenium salts added to the drinking water.

Since selenium appeared to be inversely related to some forms of cancer, attempts were made to associate demographic information and tissue concentration of selenium with the incidence of cancer. Further work was initiated to determine the effects of selenium supplementation on cancer induction in selenium-normal and selenium-deficient animals.

Identification of the chemical aspects of carcinogenesis and anticarcinogenesis and knowledge of the histochemical, morphologic, and biochemical effects of selenium have suggested possible association between optimum selenium nutrition and prophylaxis and health of biochemically stressed cells.

Neoplasia and Carcinogenesis

In 1943, Nelson et al.[540] at the U.S. Food and Drug Administration reported selenium-induced neoplasia in an experiment in which 53 of 126 principals and 14 of 18 control rats maintained on a low-protein (12%) feed lived 18-24 months. The feed of the principals contained 4.3 µg of added selenium per gram of feed in the form of seleniferous grain, potassium ammonium selenide, and potassium sulfur selenide. The mixed selenides were used as insecticides at that time and are no longer manufactured. This was the first of four series of experiments designed to induce neoplasia with selenium salts.[313,540,657,658,754,792]

The 53 principals in the Nelson experiment that lived 18-24 months had hepatic cirrhosis; 11 had hepatic adenomas, and the other 42 contained areas of adenomatus hyperplasia.[540] The 14 aged control rats had neither adenomatus and neoplastic lesions nor cirrhosis. The spontaneous incidence of hepatic tumors in the colony at that time was 0.1% in rats less than 18 months old, 0.5% in rats 18-24 months old, and 0.9% in rats more than 24 months old. Spontaneous tumors in the colony and in the principals with added selenium were associated with severe hepatic cirrhosis. These tumors were not considered to be carcinomas, and they did not metastasize.

The hepatic tumors observed in this experiment[724] probably resulted from hepatic cirrhosis rather than from the addition of selenium to the feed. Selenium was naturally present at a nutritionally adequate concentration in the control feeds, and mixed selenides and seleniferous grain were added to the feed of the principals. The cirrhosis observed in this experiment[540] has not been produced in either rats or mice maintained for their lifetime on low-protein feeds and with addition of nearly lethal concentrations of selenium salts to the feed[313,754] or water.[655] These conditions produced chronic hepatitis, but not cirrhosis.

Because of the Nelson report and interest in nutritional requirements for selenium, an extensive bioassay of selenium carcinogenesis was undertaken at Oregon State University under contract with the carcinogen-screening section of the National Cancer Institute to determine whether selenium ions would induce neoplasia in the rat.[313,754] A semipurified ration containing added selenate or selenite was fed to 1,437 rats for up to 30 months. Of these, 1,126 were autopsied when moribund and studied by histologic and hematologic methods. Those not autopsied died in cages and were too decomposed to be of value. Most of those not au-

Biologic Effects

topsied were fed 4–16 ppm of selenium and died within the first 100 days of the experiment.

Most of the 34 experimental groups contained 50–100 rats and were fed the semipurified feed. The feeds containing 8–16 μg of Se/g of semipurified feed killed rats within the first 30 days of feeding. These groups contained 15–30 rats.[380] Experimental variables included oxidation state of selenium, concentration of protein (22%, 12%, and 12% with 0.3% of methionine), added dietary selenium from 0.5 to 16 μg/g of feed, and husbandry regimens of continuous feeding or variations of partial feeding. The basal ration contained 0.1 μg of selenium per gram of feed. In addition, two groups of rats were fed a commercial ration, and two other groups received the semipurified ration with 100 or 150 μg of added N-2-fluorenylacetamide (FAA), a known hepatic carcinogen, per gram of feed.

None of the autopsied rats had hepatic cirrhosis.[380] The autopsied rats included 119 that were fed protein-deficient (12%) feed with 4, 6, or 8 μg of added selenium as selenite or selenate per gram of feed. These regimens were similar to those of Nelson and associates in protein composition (12%) and the amount of added selenium (4.3 vs. 4–8 μg/g). They differed from those used in the Nelson experiment in that the feed was composed of semipurified rather than natural feedstuffs and the added selenium had a higher valence than that used by Nelson; was not an organic compound; was not combined with ammonia, potassium, and sulfur; and was not an insecticide. Selenium as selenate was added to one group of rats fed commercial feed at the rate of 4.8 or 16 μg/g of feed. The rats fed 4 or 8 μg of selenium per gram lived more than a year and did not develop cirrhosis. Those fed the feed with 16 μg/g of added selenium died within 3 months and did not have cirrhosis.

Hepatic lesions in the principals were acute and chronic hepatitis with hyperplasia of hepatocytes.[380] The incidence of hepatic carcinoma in the rats in which it had been induced by FAA was 30%.

Acute toxic hepatitis occurred in rats fed selenium added to the semipurified ration at the rate of 4–16 μg/g and in those fed selenium added to a commercial feed at the rate of 16 μg/g.[380] These rats lived less than 100 days; were emaciated and pale; and had poor-quality coats, ascites, and edema. Chronic toxic hepatitis and hyperplastic hepatocytes were prevalent in rats fed selenium as selenite or selenate added to semipurified feeds at the rate of 0.5 or 2 μg/g for 18–30 months. Chronic toxic hepatitis was associated with bronchial pneumonia, pancreatitis, myocarditis, and nephritis. These rats did not have cirrhosis.

Six groups of rats—a total of 329 rats—were fed 0.5 or 2 μg of selenium as selenite or selenate. Of these, 276 were autopsied; 71 of those autopsied had lived 24-30 months. The longevity of these six groups was comparable with that observed by Nelson et al.[540] and Volgarev and Tscherkes.[792]

The protocol of this experiment produced lifetime exposure to concentrations of dietary selenium that ranged from nutritionally adequate (0.1 µg/g) to acutely lethal (16 µg/g). The period of selenium feeding covered the complete lifespan of all the rats fed 4-16 µg of selenium per gram—less than 280 days. In addition, 20%-30% of the rats fed the control feed or the feed supplemented with 0.5 or 2 µg of selenium per gram lived 24-30 months. Ten percent of the rats over 9 months old on the 12%-protein feed with 2 µg of added selenium per gram had small livers with regenerative nodules, as in the Nelson experiment.[540] This lesion did not occur at higher or lower rates of exposure to selenium (0.5-4 µg/g). None of the rats had cirrhosis of the liver. Harr and associates[313] concluded that the metaplasia, anaplasia, and neoplasia in the rat are not induced by selenite, by selenate, or by methionine and selenate.

Cherkes et al.[116] and Volgarev and Tscherkes[792] reported the third experiment on the carcinogenic potential of selenium. In three series of experiments, they fed selenium as selenate to 200 rats at the rate of 4.3 or 8.6 µg/g of feed. The feeds were not semipurified. They contained 12%-30% protein with addition of riboflavin, methionine, α-tocopherol, cystine, nicotinic acid, and choline in appropriate groups. Groups of rats without selenium supplementation (controls) were not included in the experiments.

In the first series of 40 rats, 23 lived 18 months or longer. All 40 rats had cirrhosis, as reported by Nelson et al.[540] Of these, 4 developed sarcomas, 3 developed hepatic carcinoma (2 with metastases), and 3 had hepatic adenomas.[116,792] Of the 13 noncancerous rats, 4 had lesions termed precancerous. The lesions were cholangiofibrosis, oval cell (bile duct cell) proliferations, and biliary cysts.

In the second series, 60 rats were fed selenium and 12% or 30% casein.[116,792] One liver carcinoma, one hepatic adenoma, and three sarcomas were reported. In the third series, none of the 100 rats had neoplasms or precancerous lesions.

The 15 neoplasms observed in the three series occurred in about 120 rats that lived 18-24 months.[116,792] The 15 neoplasms included 7 sarcomas, 4 hepatic carcinomas, and 4 hepatic adenomas. Six of the sarcomas appeared to be extrahepatic lymphomas, and the seventh appeared to have arisen from the bile ducts or mammary glands.

Six of the eight liver cancers, all four of the hepatic "precancerous" lesions, and four of the seven sarcomas occurred in the first series.[116,792] Two liver neoplasms and three carcinomas occurred in the second series, and none occurred in the third series. Thus, the incidence of hepatic neoplasia in the first series was 40% of the old rats; in the second series, 12%; and in the third series, none.

The fourth experiment on the bioassay of selenium carcinogenesis was reported by Schroeder[655] and Schroeder and Mitchener.[657,658] Groups of 100-110 rats and mice were supplemented with 2 or 3 ppm of selenium as selenate or selenite in the drinking water. The 90% survival time of rats supplemented with selenate was 1,113 days compared with 1,180 for the nonsupplemented rats. Selenium-supplemented rats were 3%-7% heavier than the nonsupplemented controls after 12-36 months of supplementation. Despite heavy supplementation with selenium salts at near-lethal amounts (about 25 times the amounts for the controls), the concentration of selenium in kidneys, liver, heart, lungs, spleen, and erythrocytes of selenium-supplemented rats was 1.5-2 times the concentration in tissue from control rats.

Mice given selenite or selenate and rats given selenite through the drinking water did not have an increased incidence of tumors.[655,657,658] About 70% of the rats were autopsied, and 65% of those autopsied were examined histologically. Neither the rats nor the mice had hepatic cirrhosis. However, a severe epidemic of chronic murine pneumonia occurred in the rats. Twenty tumors were found in the control group of rats, and 30 were reported in the selenate-supplemented group. The types of tumors and the numbers were given as follows:

	Control Group	Selenate Group
Mammary tumors	10	11
Spindle cell sarcomas	2	4
Leukemia types	2	4
Pleomorphic carcinomas	1	2
Other sarcomas	5	11

The anatomical location of the sarcomas and pleomorphic carcinomas was not reported. These tumors may have been sclerosed granulomas secondary to the epidemic of chronic murine pneumonia. Histologic slides were prepared only from selected organs and animals, and the rationale for that selection was not reported. Since the organs and tissues were not systematically searched, type and incidence of histologic lesions are not known.

Despite an initial report of selenium as a carcinogen,[540] chronic experimental exposure of rats and mice to selenium salts over a period of 12 years has not induced neoplasia.[116,313,540,655,657,658,680,754,792] During the same period, selenium salts have been used prophylactically and therapeutically in ruminants, omnivores, and carnivores throughout the world.

Epidemiologic and demographic evidence from the widespread use of selenium supplementation, exposure to toxic concentrations of selenium in feeds, and use of selenium in shampoos and industrial plants does not suggest that selenium is carcinogenic; rather, it may be correlated with a reduction in the evidence of human and ovine cancer.[224,657,673,681-684,807]

There has been no increase in the incidence of neoplasia in any of the treated species. In New Zealand, the incidence of intestinal cancer in treated sheep has decreased.[705]

Some regions of the world, including the North Central and Rocky Mountain regions of the United States, are geologically rich in selenium. Forage plants in these regions often contain more than 10 µg of selenium per gram of weight. Ruminants and horses eating these plants develop selenosis and may die. Shamberger[680,684] reported a lower incidence of cancer in people living in these regions. This observation was correlated with lower-than-average concentrations of selenium in the blood of some patients with cancer.[682]

Anticarcinogenesis

The demonstration of the relationship of selenium to human cancer is limited to demographic studies and comparisons of blood levels of selenium in patients with and without malignancies. However, Weisberger and Suhrland[809] discussed the effect of selenium cystine on leukemia, and Chu and Davidson[121] listed selenium compounds among potential antitumor agents.

Demographic and experimental observations of Shamberger and associates support the concept of pharmacologic and medical uses of selenium salts.[653,680,682,684] They found an inverse correlation between the incidence of cancer deaths, the concentration of selenium in the patient's serum, and the geographic incidence of selenium—low, moderate, or high. The concentration of selenium in the blood of cancer patients averaged 74% of normal. However, the blood of patients with some forms of cancer contained normal concentrations of selenium.

The selenium contents of diets of 17 paired human males with and without gastric cancer were compared and related to dietary antioxidants and food preservatives.[683] Patients with gastrointestinal cancer or metastases to gastrointestinal organs had significantly lower levels of selenium in the blood than normal patients.[683] No elevations of selenium in the blood of cancer patients were noted. The authors postulated that selenium acted to prevent attachment of the carcinogen to DNA sites.

Shamberger[717] also reported on the effect of adding sodium selenide to cancer-inducing preparations of anthracene compounds or adding sodium selenite to the feed of rats exposed to anthracene compounds. Rats fed dietary selenite and those treated with preparations of anthracene compounds with added selenide developed fewer skin papillomas than rats treated with anthracene compounds without added selenide.

Harr et al.[315] reported that after 200 days of feeding selenium-depleted rats a semipurified feed containing 100 ng of the hepatic carcinogen FAA per gram of feed and 0.1, 0.5, or 2.5 μg of added selenium as selenite per gram, the incidence of mammary and hepatic neoplasia with or without 0.1 μg of added selenium per gram was three times greater than the incidence in rats supplemented with 0.5 or 2.5 μg of selenium per gram. The low-selenium groups (0 and 0.1 ppm) died before 200 days of age and had a 90% incidence of neoplasia. At this time, 35% of the rats fed 0.5 and 2.5 μg of selenium per gram had died, and the incidence of neoplasia was 30%. Most of the remaining rats fed 0.5 and 2.5 μg of selenium per gram lived for an additional 120 days. By this time, they had received the carcinogen for an additional 120 days, and the total incidence of neoplasia was 90%, as observed in the groups receiving 0 and 0.1 μg of selenium per gram. Since the longevity of the rats was proportional to the amount of selenium supplementation and the duration of exposure to the carcinogen, the increase in cancer in the rats heavily supplemented with selenium may have been due to greater exposure to the carcinogen or to longer time for induction.

The mammary tumors in the group not supplemented with selenium were more invasive than those in rats from the three supplemented groups and predominated in the pelvic rather than in the thoracic region, as in the selenium-supplemented or commercially fed rats.

Johnston[380] studied the effect of selenium on the induction of cancer by FAA and diethylnitrosamine in selenium-depleted rats over a restricted exposure period. Because of widely varying rates of feed consumption by the principal and control groups and the high incidence of neoplasia in all the exposed groups, results were confusing.

The unique ability of selenium to reduce methylene blue was reported by Schrauzer and Rhead,[653] who suggested that this ability might provide a basis for testing for cancer or susceptibility thereto. Clayton and Baumann have reported on the relation between diet and azo dye tumors.[123] In studies of lipid therapy based on the types of lipid imbalance in cancer patients, Revici[622] reported that the most satisfactory and reproducible palliative effects were obtained by using synthetic lipids containing bivalent selenium, a serendipitously acquired observation alluded to by Frost.[224] Berenshtein and Aleshko have described the effect of selenium on lipid metabolism.[45]

Clinical observation of the efficacy of parenterally administering mixtures of selenium and vitamin E to animals with adequate dietary selenium and vitamin E indicates that an additional 1–3 mg/50 kg of body weight per month improves functioning of reproductive, muscular, and vascular systems. Efficacy is claimed in clinical cases of stiffness, lameness, myositis, hepatic degeneration, infertility, loss of condition of the hair coat, and poor growth. In lameness and stiffness in race horses, dogs, breeding rams, and bulls, a severe strain is put on the musculoskeletal system and in particular on the animal's joints. Veterinarians in selenium-deficient areas believe that animals improve after receiving selenium supplements. Theoretically, this could be related to better vascularization of the tissues or to the role of glutathione peroxidase in maintenance of cellular membranes.

PHYSIOLOGIC ROLE

Vitamin E

The close nutritional interrelationship between vitamin E and selenium suggested that selenium might function *in vivo* as an antioxidant.[743,744] Five general mechanisms were postulated to account for the hypothetic antioxidant activity of selenium compounds: peroxidation inhibition, peroxide decomposition, free radical scavenging, repair of molecular damage sites, and catalysis of protector sulfhydryl compounds. Although many of the *in vitro* and *in vivo* effects of selenium can be rationalized by an antioxidant action of selenium, a number of recent studies indicate that selenium probably has a more subtle role *in vivo*. For example, dietary selenium had no influence on the survival of rats exposed to chronic whole-body irradiation.[362] Moreover, selenium had no protective effect against the lipid peroxidation induced in mitochondria by ascorbate, oxidized plus reduced glutathione, or iron.[435] Finally, there are several reports that show beneficial responses to selenium even in animals adequately supplied with vitamin E.[480,724,750]

Glutathione Peroxidase

The observation of Rotruck *et al.*[643] that dietary selenium could protect erythrocytes against the oxidative hemolysis characteristic of vitamin E deficiency as long as glucose was present in the incubation medium suggested an involvement of selenium in glutathione (GSH) metabolism. Since the level of GSH in red blood cells was the same in selenium-deficient animals as in selenium-adequate animals,[644] it appeared that selenium deficiency had no effect on generation of GSH. Rather, there seemed to be a fault in the utilization of GSH in selenium deficiency, and, indeed, the enzyme GSH peroxidase has recently been shown to contain selenium.[206,645] This discovery of a role for selenium in GSH peroxidase is highly significant in that it is the first demonstration of a role for selenium in a specific mammalian enzyme.

Nonheme Iron Proteins

Diplock and co-workers first elaborated the hypothesis that the biologically active form of selenium may be selenide in the active site of an uncharacterized class of nonheme iron proteins.[111,157,158,450] This concept was based on the finding that significant portions of the selenium in the mitochondrial and microsomal fractions of rat liver were in the selenide valence state in animals with adequate vitamin E. However, in animals fed a diet deficient in vitamin E, little selenide was detected in the subcellular fractions. Therefore, vitamin E was considered to protect the unstable selenide from oxidation. This theory is appealing, because, if selenium truly has a role in the active site of a mitochondrial nonheme iron protein, this might explain the decline in respiration that is characteristic of liver slices prepared from rats fed diets deficient in both vitamin E and selenium.[117]

Electron Transport

Evidence that selenium plays a role in the electron transport chain was recently presented by Levander and colleagues.[434] This idea was derived from experiments that examined the effect of dietary vitamin E or selenium on the swelling of rat liver mitochondria induced by various chemical agents added *in vitro*. Previous work had shown that dietary vitamin E, but not selenium, was able to protect mitochondria against the swelling caused by compounds that promoted lipid peroxidation.[435] Selenium, on the other hand, accelerated the swelling caused by certain thiols.[433] Studies with respiratory inhibitors indicated that the swelling caused by the addition of GSH plus selenite *in vitro* might be partly mediated at the level of cytochrome c.

Selenium was then demonstrated to be a highly effective catalyst for the reduction of cytochrome c by GSH in a chemically defined model system.[434] It was suggested that selenium may function *in vivo* by facilitating the transfer of electrons from GSH or other sulfur compounds into the cytochrome system. A possible role for selenium in biologic electron-transfer reactions is also supported by the work of Turner and

Stadtman,[766] who found that a selenoprotein was a component of the clostridial glycine reductase system. Moreover, Whanger et al.[819] have found a selenoprotein in lamb muscle that contains a heme group identical to that of cytochrome c.

Interrelationships

Although each of the above hypotheses is a distinctive way of considering the role of selenium in living systems, all the theories are closely related, and it may well turn out that each one is just a variation on a basic general theme. For example, many of the "antioxidant" properties of selenium could be explained by a role for selenium in GSH peroxidase, since this enzyme destroys peroxides. Also, GSH peroxidase has been shown to be involved in mitochondrial swelling and contraction.[543] Obviously, any role for selenium in mitochondrial nonheme iron proteins would be intimately related to electron transfer. Thus, it is reasonable to suppose that one underlying action of selenium could account for all these phenomena. Whether this action of selenium is related to its role in GSH peroxidase is a question that can be answered only by additional research.

MEDICAL USES OF SELENIUM IN HUMAN BEINGS

This section is concerned with the medical uses of selenium in human beings. These uses fall into two categories: (1) the diagnostic, which uses the radionuclide form, ^{75}Se, usually as selenomethionine, for scanning of organs and tissues, and (2) the therapeutic, which uses selenium sulfide for the treatment of seborrheic dermatitis and tinea versicolor.

Diagnostic Scanning and Labeling

^{75}Se-selenomethionine was the first amino acid used in clinical scintillation scanning. It can be produced with a high specific activity by biosynthesis (about 300 mCi/mg of selenomethionine) or chemical synthesis (6 mCi/mg).[26] This nuclide has several gamma emissions with sufficient tissue penetration to allow external clinical measurements of nuclide *in situ*.

PANCREAS

The high rate of incorporation of amino acids into pancreatic enzymes and the similarity of metabolism of selenomethionine and methionine formed the basis for the suggestion by Blau[56] of the use of ^{75}Se-selenomethionine for external visualization of the pancreas, an organ impossible to delineate by X ray. Clinical trials confirmed that ^{75}Se-selenomethionine accumulated in the pancreas in relatively large quantities[57,58] and in concentrations eight to nine times greater than in the liver and small bowel 1 hr after intravenous administration.[730]

Pancreatic scintigraphy has been used for determining such pathologic conditions as pancreatic carcinoma, pseudocyst, and pancreatitis. Although there has been some disagreement as to the overall value of this procedure, many authors have endorsed it, with varying reservations, as a useful addition to the armamentarium available for investigation of a diagnostically refractory organ.[73,172,632] Eaton et al.[173] stress the potential value of ^{75}Se-selenomethionine imaging as a means of documenting the return of pancreatic function to normal after an attack of acute pancreatitis or after corrective surgery for chronic pancreatitis. Agnew et al.[4] performed a different test of pancreatic function simultaneously with the scan by collecting duodenal samples of the labeled digestive enzymes secreted by the pancreas. Following the stimulus of a test meal, normal patients showed an increase of ^{75}Se that correlated well with the trypsin concentration in the duodenal aspirate. There was no increase in ^{75}Se in the duodenal aspirates of patients with carcinoma of the pancreas or chronic pancreatitis.

LIVER

Scintigraphic imagery of the liver usually involves scanning with radiogold, ^{99m}Tc sulfur colloid, or ^{113m}In ferric hydroxide, all of which are deposited in the reticuloendothelial system. The principal parenchymal cell label is ^{131}I rose bengal. A focal defect in the liver is defined classically by its inability to concentrate the above-labeled substances from the vascular system. In the focal lesion, the incorporation of ^{75}Se-selenomethionine is related to its vascularity and capacity to metabolize methionine. Modification of the rectilinear scanner permits it to function as a dual-channel subtraction mode to subtract ^{198}Au from ^{75}Se in the liver and to present the display of the liver in one color and the selenium in the pancreas and the focal lesion in another. The avascular lesions, including cysts, abscesses, scars, pseudotumors, and extrinsic pressure defects, show little or no ^{75}Se activity. Metastatic lesions may have varying levels of selenium concentration. The hepatocellular carcinoma concentrates selenium in amounts equal to the normal hepatic parenchyma and can only be visualized in the ^{75}Se-selenomethionine scan by the subtraction technique.[106,174,385] Stolzenberg[730] concludes that if melanoma is clinically excluded, hepatocellular hepatoma can be strongly suggested by ^{75}Se-selenomethionine liver scanning when an area of defect on colloidal scan shows activity on ^{75}Se-selenomethionine scan.

PARATHYROID, THYROID

Potchen[614] and later others[154,325] showed that localization of ^{75}Se-selenomethionine in the parathyroid gland was sufficiently higher than in the suppressed thyroid gland and other surrounding tissues to make it usable for identification of parathyroid adenomas. Ashkar et al.[25] claim better parathyroid visualization following the administration of glucagon, an effective stimulant to incorporation of ^{75}Se-selenomethionine. Thomas et al.[748] differentiated malignant from benign lesions of the thyroid gland by using complementary scanning with ^{75}Se-selenomethionine and radioiodide.

LYMPHOMAS AND MISCELLANEOUS TUMOR MASSES

Herrera et al.[331] reported that ^{75}Se-selenomethionine given intravenously for pancreatic scanning was incorporated into lymphomas in sufficient amounts to be detected by external means. Spencer et al.[722] confirmed the avidity of lymphomas for selenomethionine. On further investigation of patients with nonneoplastic diseases and various abdominal tumors, it was found that neither normal lymph nodes nor the common epithelial neoplasms accumulated enough activity to be clearly imaged externally.[330] The studies of Ferrucci et al.[197] concluded that a negative isotope study did not reliably exclude disease of lymphatic tissue, whereas an abnormal scintigram indicated a 70%–80% likelihood that disease was present.

Goal et al.[260] point out that ^{75}Se-selenomethionine scanning is a simple, safe, and atraumatic procedure for visualizing mediastinal masses and therefore is preferable to other investigative procedures, such as pneumomediastinography and venography. They found this radiopharmaceutical incorporated in actively dividing cells of various neoplasms, including bronchogenic carcinoma, and in occult thymomas. In an attempt to exploit for diagnostic purposes the ability of neuroblastoma cells to synthesize cystathionine, D'Angio et al.[142] confirmed localization of ^{75}Se-selenomethionine within the tumors in several patients.

PLACENTAL COMPETENCE

Because the methods for measuring intrauterine growth rate were considered unsatisfactory, Garrow and Douglas[245] suggested the measurement of placental competence by the administration of ^{75}Se-selenomethionine to the mother, which would assess one of the more important functions of the placenta: its ability to take up amino acid from the maternal circulation and pass it to the fetus. The high growth rate of the fetus can be achieved only if the placenta transmits nutrients to the fetus in sufficient quantities; some nutrients, notably amino acids, have to be transported from the maternal to the fetal circulation against a concentration gradient. If the placenta is unable to support the fetus in this way, intrauterine growth is retarded, and the baby, when delivered, is "light-for-dates"; such babies have high morbidity and mortality rates.[245]

Mothers who had low rates of transfer of ^{75}Se-selenomethionine had significantly smaller babies than those with normal transfer rates, and high transfer rates tend to be associated with rather big babies. The test can be done quickly and easily, and it seems to give reliable results at any time after 28 weeks of gestation, but it is probably most useful at about 34 weeks.[159] Although the test involves giving a dose of radioactivity to a pregnant woman, it cannot be considered dangerous, according to Douglas;[159] the radiation involved is about 3% of that given by X-ray examination of the abdomen and about one ten-thousandth of that which has been known to harm a fetus at this stage of gestation. In performing this test on 467 patients, Lee and Garrow[423] found no adverse reactions attributable to the test.

EXTRACELLULAR FLUID VOLUME, PLATELETS, FIBRINOGEN

Albert et al.[5] suggested the use of ^{75}Se sodium selenate instead of ^{35}S sodium sulfate for the measurement of extracellular fluid volume. They stated that the advantages of ^{75}Se over ^{35}S are the simplicity of detecting a radioactive element, such as ^{75}Se, having only gamma emissions; the convenience of accurately measuring ^{75}Se with commercially available and moderately priced equipment; and the fact that ^{75}Se simplifies work in studies where two or more tracers are used. However, since selenate and sulfate are not metabolically equivalent, the substitution of ^{75}Se for ^{35}S is not recommended.

The incorporation of intravenously injected ^{75}Se-selenomethionine into platelets was found to vary with alterations in the rate of platelet production. Penington[601] concluded that ^{75}Se appears to label newly produced platelets during their formation in megakaryocytes and provides a method by which thrombopoiesis may be studied *in vivo*. In thrombocytopenic states, ^{75}Se-selenomethionine is an effective and clinically useful cohort label of platelets and fibrinogen.[67]

Therapeutic Uses

Because of the close resemblance of selenium to sulfur and the fact that some of the most powerful synthetic drugs contain sulfur, efforts have been expended in synthesizing and testing organic selenium compounds as potential chemotherapeutic agents. Klayman[395] reviewed the attempts at chemotherapy with organic selenium compounds, concluding that few of them offer enough advantages over present agents to warrant clinical trial. Therapeutic uses of selenium in human beings have been limited to external application of selenium sulfide preparations in dermatologic disorders of seborrheic dermatitis and tinea versicolor.

SEBORRHEIC DERMATITIS

Selsun is a commercial preparation of selenium sulfide in aqueous suspension containing emulsifying, buffering, and carrying agents for use as a shampoo in cases of dandruff. A prescription product (Selsun Red) contains 2.5% (w/v) selenium sulfide; a product sold over the counter without a prescription (Selsun Blue) contains 1% (w/v) selenium sulfide.

In one of the earliest investigations on the subject, Slinger and Hubbard[708] reported that more than 80% of the cases of seborrheic dermatitis of the scalp treated with selenium sulfide (2.5%) were controlled completely during the period of use of the shampoo without observable signs of cutaneous irritation, sensitization, or toxicity. Slepyan[707] and Sauer[648] reported essentially the same results. Orentreich et al.[574] found that the 2.5% preparation was as effective against seborrheic dermatitis as another popular shampoo containing zinc pyrithione as the active ingredient.

Some investigators noted side effects from the use of selenium sulfide (2.5%) shampoos. Eisenberg[179] described three cases of contact dermatitis. Increased sebum secretion[262] and oiliness of scalp[47,574] have been observed. Bereston[47] also reported an orange tinting of gray hair. Diffuse loss of hair, which ceased upon discontinuation of the shampoo, was observed by Grover[273] and Sidi and Bourgeois-Spinasse.[703] Although Archer and Luell[20] observed dysplastic changes in the hair roots of persons using selenium sulfide suspensions, they called for experimentation with other sulfides and for control studies of shampoos consisting only of emulsifying, buffering, and carrying agents. In acute and chronic applications of selenium sulfide shampoo, Maguire and Kligman[457] concluded that there were no root deformities or changes in rate of hair regrowth.

Ayres and Ayres[29] found 1%-selenium-sulfide ointment to be effective in treating seborrheic dermatitis of glabrous skin in 73% of patients tested, but a 21% incidence of irritation and positive patch test reactions led to the substitution of a 0.5% ointment that caused less irritation and was still effective.

Ransone et al.[619] reported a case of systemic selenium toxicity in a woman who had been shampooing two or three times weekly for 8 months with selenium sulfide suspension. Although the intact skin ordinarily absorbs little, if any, selenium sulfide, the presence of open scalp lesions on this patient were considered to be instrumental in permitting sufficient systemic absorption to cause systemic toxicity in the form of progressive tremor, weakness, and anorexia. Symptoms cleared rapidly when use of the shampoo was discontinued.

TINEA VERSICOLOR

Selenium sulfide has been used in treating tinea versicolor since 1953,[168] but treatment procedures are varied,[250,429,629] and follow-up studies of the effect are scanty. Albright and Hitch[6] found that a single overnight treatment was promising as an easy form of suppressive remedy for this fungal infection, but advocated repetition of the treatment three or four times at least every third month. He noted no manifestation of skin irritation in his patient series, although the genital region was avoided. Hersle[332] concludes that the risk of intoxication with water-soluble selenium sulfide by this mode of application is negligible in view of studies by Henschler and Kirschner[328] showing low toxicity and low absorption rates, not enhanced by simultaneous administration of detergents.

6

Sampling and Analysis

COLLECTION, PREPARATION, AND STORAGE OF SAMPLES

Because of the volatile nature of many selenium compounds, several difficulties are encountered in preparing, storing, and analyzing certain types of materials. In addition, various plants and animals may differ considerably in selenium content, soils may differ greatly in selenium content at different depths, waters contain both soluble and suspended forms of the element, and air contains both gaseous and particulate forms. Thus, obtaining representative samples for analysis is sometimes difficult. Finally, the method of preparing samples for analysis, often dictated by the method of analysis that is going to be used, can introduce errors into the results. Consideration of these matters is in some cases as important as consideration of the method used to make the analysis.

Plants

In sampling plants, it should be kept in mind that leaves, roots, stems, and seeds often differ considerably from one another in selenium content.[41] Their tissues are normally dried at 70° C or less for analysis, since some higher temperatures have been reported to cause large selenium losses.[503] For most samples, the loss of the element at this temperature is probably very small;[24,178] but some of the indicator plants, such as *Astragalus bisulcatus*, contain significant amounts of volatile selenium compounds,[40] and these should be analyzed without drying. Leaves reduce to fine particles much more readily than do stems, and the tendency of these plant parts to remain separated during grinding and subsequent handling increases the possibility of sampling error.[10] Thus, fine grinding and complete mixing are essential to accurate results. In view of reports suggesting selenium losses during long-term, open storage without temperature control,[212,512] samples of dry plant material should be stored in tightly stoppered bottles at 0°–5° C. Undried samples of tissue should be stored frozen in closed containers.

Animal Tissues

Animal tissues should also be stored frozen. Blood should be separated into plasma or serum and cell fractions prior to freezing if separate values for them are desired. This reduces the likelihood of enzymatic formation of volatile selenium compounds. It also reduces the likelihood of significant loss of the element from urine samples.[191] When drying for analysis is necessary, it should be accomplished by lyophilizing.

When wet digestion procedures are used in analyzing plant or animal tissue, some sampling error can be avoided by predigesting a large sample in nitric acid and then using an aliquot for completion of the analysis.[560]

Water

Water should be sampled by commonly accepted methods. If it contains suspended material, this should be removed by filtration within a few hours, preferably immediately after the sample is collected. The selenium content of the sediment can be determined separately, if it is essential that this be known. The filtered sample should be stored at 0°–5° C to reduce the possibility of microbial action that could result in the precipitation or removal of selenium. Usually, an appropriate amount of the sample is made alkaline to phenolphthalein and then evaporated to a small volume or to dryness for analysis. This assumes that no forms of the element are present that would be volatile under these conditions. Normally, this assumption can be made, but additional studies on ways of reducing volume seem needed.

Air

Little study has been given to air sampling for selenium analysis. Most of the work reported to date has dealt only with the particulate matter that can be removed by filters. Tabor et al.[741] recently published a tentative method for analyzing this material. However, Pillay and Thomas[607] found selenium to be divided into gaseous and suspended forms, and improvements in collecting the gaseous form are needed.

METHODS OF ANALYSIS

Detailed discussions of methods of selenium analysis have been presented elsewhere,[571,802] and the subject will be only briefly reviewed here. Many procedures are available for use, and these will be discussed in general terms without attempting to completely cover the literature. They may be divided into two general classes: those that do not require the destruction of the sample (nondestructive) and those that require getting rid of the organic matter before the selenium is measured (destructive).

Nondestructive Analysis

X-RAY FLUORESCENCE

Nondestructive analysis can be accomplished by X-ray fluorescence analysis,[261,307,446,559] but this method at present lacks sensitivity and has not been widely used.

NEUTRON ACTIVATION ANALYSIS

On the other hand, neutron activation analysis (NAA) can be used either with or without destruction of the sample, and it has become increasingly popular in the latter use. When used with destruction of the sample, the selenium is usually separated from interfering elements following activation and prior to its measurement. With or without destruction, the method has been used on a wide variety of samples, some examples being lungs,[617] muscle,[268] serum,[501] blood,[64] kidneys and liver,[419] pancreas,[538] hair,[48] eye lens,[199] dental enamel,[551] feces,[788] yeast,[665] tobacco,[530] forages,[341] feeds,[299] wheat flour,[531] cystine,[668] rainwater,[618] air,[607] and fossil fuels.[608]

The chief disadvantage of NAA for selenium analysis is the cost of the equipment. This is so great that it can hardly be justified for selenium

Sampling and Analysis

analysis only. In survey types of work where multielement analysis is conducted on large numbers of samples, it may be the method of choice, since it can greatly speed the work (when used nondestructively) and since it has excellent sensitivity for a number of elements. There is evidence, however, that at submicrogram levels of selenium the analysis of natural materials by this method can yield inaccurate results,[571] so the method should probably be tested in each laboratory on a variety of samples of known selenium content before it is used routinely.

Destructive Analysis

DESTRUCTIVE METHODS

Other methods of analysis require the destruction of organic matter before measurement of the selenium. This destruction has been accomplished in a variety of ways: dry ashing with magnesium nitrate;[631] low-temperature ashing with excited oxygen;[255] closed-system combustion with the Schöniger oxygen flask[286] or with the Parr bomb;[170] wet digestion with sulfuric acid,[631] nitric and sulfuric acid,[829] mixtures of nitric and perchloric acids alone[803] or with sulfuric acid,[120] ammonium vanadate,[482] sodium molybdate,[139] or hydrogen peroxide;[298] and hydrogenolysis.[752] Of these, wet digestion with mixtures containing perchloric acid is most commonly used, although oxygen-flask combustion is also popular. Wet digestion has the advantage of being easily adaptable to large numbers of samples and to liquids or materials of high moisture content.

SEPARATION FROM INTERFERING SUBSTANCES

Some methods call for separating the selenium from interfering substances before measuring it. This can be accomplished by reduction of the selenium to its elemental form from filtered digests and then removal by filtration;[62] distillation as the tetrabromide with[631] or without[308] reduction to elemental selenium; extraction with an ethylene chloride–carbon tetrachloride mixture of the complex of the element with toluene-3,4-dithiol, followed by wet digestion and completion of the analysis;[804] arsenic coprecipitation;[452] and ion exchange treatment to remove interfering cations.[662] Although distillation and precipitation of the element has been commonly used in the past, arsenic coprecipitation is simpler and much more satisfactory for small amounts of selenium, and it is now the most commonly used method.

SPECTROPHOTOMETRY, GAS CHROMATOGRAPHY, ATOMIC-ABSORPTION SPECTROMETRY, POLAROGRAPHY, OXIDATION-REDUCTION TITRATION, SPARK-SOURCE MASS SPECTROMETRY, NEUTRON-ACTIVATION ANALYSIS, AND FLUOROMETRY

Measuring selenium in digests or combusted materials, in some cases with and in others without its separation, as discussed above, has been accomplished by precipitating the selenium in elemental form as a pink sol and estimating the amount visually by comparison with standards;[129] filtering the precipitated element on a barium sulfate pad[209] or millipore filter[544] and comparing visually with standards; the ring oven technique, in which 3,3'-diaminobenzidene (DAB) is used to develop the color, and comparing with standards;[812] spectrophotometric determination of the product of reaction with DAB,[348] 2,3-diaminonaphthalene (DAN),[449] other o-diamines,[114,136,150,649,752] and 2-mercaptobenzothiazole;[44] 2-mercaptobenzoic acid[134] methods based on the catalytic effects of selenium[388,813] or the production of an intermediate that is then reacted to give an indirect measure of the element;[134,394,533,577] gas chromatography;[534] atomic-absorption spectrometry;[384] polarography;[120] oxidation-reduction titration;[86,152,397,717] gravimetric determination of the precipitated element[503] or the insoluble Se-DAN complex;[449] spark-source mass spectrometry;[318] neutron-activation analysis (see previous discussion); and the fluorometric determination of the Se-DAB complex[130] or the Se-DAN complex.[591] Of these, the measurement of the fluorescence of the Se-DAN complex and neutron-activation analysis are most commonly used today, and both are capable of determining submicrogram levels of the element. Many satisfactory procedures, differing only slightly from one another, are available for those choosing the first of these commonly used methods. Where the equipment is available, and where multielement survey work is being done, spark-source mass spectrometry will probably be used more frequently. Finally, with improvements in instrumentation and procedures, atomic-absorption spectrometry could become the method of choice in many laboratories.

Regardless of the method selected for use, it should be recognized that all are subject to error, and each laboratory should test its procedures by analyzing—preferably by more than one method—a variety of samples of known selenium content. A recent report on a comparison of methods for fossil fuel analysis supports this conclusion.[793]

7

Summary and Conclusions

SUMMARY

Chemistry

The chemical properties of selenium are similar to those of sulfur. It exists in nature in several oxidation states: −2, 0, +4, and +6.

In its −2 state, it occurs as hydrogen selenide, a highly toxic and reactive gas that decomposes quickly in the presence of oxygen to elemental selenium and water. Heavy metal selenides are insoluble, and a number of organic selenides having properties similar to those of organic sulfides have been identified in biologic materials. Some of these are very volatile.

In elemental form, selenium is insoluble and not subject to rapid oxidation or reduction in nature. Because of its insolubility, it is not toxic. On burning, it is oxidized to selenium dioxide, which sublimes and, on solution in water, forms selenious acid.

Selenium occurs in the +4 state as inorganic selenites. In soluble form, these are highly toxic. Selenite has an affinity for iron and aluminum sesquioxides, with which it forms stable adsorption complexes. This and the ease of selenite reduction to elemental selenium under acid and reducing conditions make this form quite unavailable to plants and also reduce the probability of pollution of water by the element.

Alkaline and oxidizing conditions favor the formation and stability of the +6 form of the element, selenate. Most selenates are quite soluble and highly toxic. This form of the element is not tightly complexed by sesquioxides. In soils, selenates are easily leached and are available to plants.

Biologic processes appear to be involved in reduction of the element. Oxidation apparently occurs in alkaline soils by chemical weathering, and reduction also results from burning. The reduction process can produce volatile organic selenides or hydrogen selenide; burning can produce particulate elemental selenium or selenium dioxide, and these are the forms most likely to occur in the atmosphere.

Occurrence

The earth's crust is estimated to contain an average of about 0.1 ppm of selenium. This is not evenly distributed in the lithosphere. Some mineral deposits of limited size contain over 20% of the element; some rocks, undetectable traces.

Chemical, and possibly microbiologic, oxidation in alkaline soils solubilizes selenium as selenate, making it available to plants. In acid soils, the element exists in the more reduced forms, which are not very available to plants. Thus, two factors have been of major importance in the development of soils that produce crops containing too little or too much of the element: the selenium content of the parent materials and the conditions of pH under which the soils are formed. To a large extent, the latter is related to rainfall, and in the United States the areas of excessive selenium are the more arid ones. As rainfall increases, the probability of selenium deficiency increases.

With possible exceptions in the case of fish and certain seafoods, selenium enters the food chain almost entirely via plants. Although some edible animal tissues, especially liver, tend to accumulate the element, there is no serious food chain buildup. Our present food habits, food-processing methods, and transportation capabilities seem to preclude the possibility of too much or too little selenium in man's diet in the United States.

Waters seldom contain significant concentrations of the element. Indeed, only in rare cases have they been found to contain enough to be considered a good source for supplying the body with adequate nutritional levels.

Data on the selenium content of fossil fuels are limited, but coals containing a few parts per million are apparently not unusual. Fuel oils

Summary and Conclusions

probably contain lower concentrations of the element. During combustion, some of the selenium in these fuels can escape into the atmosphere.

Industrial and Agricultural Uses

The free world refinery production of selenium from 1964 through 1971 averaged 2.2 million lb. The United States was the leading producer for most of those years, followed by Canada, Japan, and Sweden. Nearly all primary selenium production derives from copper ores. Recovery is by treatment of residue slimes generated during electrolytic refining of copper.

Apparent annual domestic consumption of selenium in recent years approximated 1 million lb. Electronic applications, including use in rectifiers, xerographic copying machines, and photoelectric cells, account for a substantial part of selenium consumption. Selenium is used in the glass and ceramics industry, in the manufacture of pigments, as a component of plating solutions, and as a chemical agent in the preparation of many products.

Selenium was once used in controlling insects on ornamental plants, but this is no longer done. At present, the chief use of the element in agriculture is in the prevention of selenium deficiency in livestock and poultry.

Cycling

Like other elements, selenium is being continuously cycled by natural processes. Qualitative aspects of the cycling have been well documented, and the various pathways followed among waters, land, air, and living organisms can be stated with considerable accuracy. However, quantitative data are meager, and we are not sure about the importance of the various pathways.

Few data are available for direct determination of industrial atmospheric emissions of selenium. However, it was estimated that 2.4 million lb of selenium was discharged by industrial plants in 1970. Burning of coal accounted for 62% of the total, and production of copper for 26%. The remainder was about equally divided among selenium-recovery plants, glass manufacturing, and the burning of fuel oil.

If estimates are correct, the amounts of selenium emitted from industrial sources into the air are small in comparison with the amounts of other industrial pollutants emitted; and if adequate dispersal procedures are followed, it seems that selenium emissions should not present a serious problem. Industrial emissions of selenium probably occur as finely divided solid particulates.

Biologic Effects

Plants contain widely varying amounts of selenium, the amounts depending on the plant species, the amount and form of the element in the soil, and other factors. An essential role for selenium has not been established in plants, and the element is not known to have toxic effects on plants under natural conditions. In many respects, plants metabolize selenium as they do sulfur, but significant differences have been noted in the biochemical pathways of these elements.

The absorption of orally ingested, soluble selenium compounds appears to be virtually complete in monogastric animals. In ruminants, some formation of insoluble elemental selenium may occur that would decrease the absorption of selenium. At least in the case of selenite, there seems to be no mechanism to regulate the amount of selenium absorbed by the gastrointestinal tract. Little quantitative information is available regarding the absorption of selenium compounds through the lungs or skin.

The distribution of ingested selenium is widespread in the internal organs, the largest amounts occurring in the liver and kidneys. When very small doses of selenium are given, the testes retain a sizable fraction of the dose given in long-term studies. Otherwise, the pattern of selenium distribution in the organs is generally similar, regardless of whether toxic or physiologic doses of selenium are used.

The main excretory pathway for selenium in monogastric animals is the urine, and the amount excreted via this pathway is directly related to the amount of selenium in the diet. The amount that monogastric animals excrete in the feces is minor, but the amount that ruminants excrete by this route can assume major importance. Excretion of volatile selenium compounds via the lungs becomes important only when animals are exposed to toxic levels of the element. Bile, pancreatic juice, saliva, and hair normally represent insignificant routes of selenium excretion.

Organic forms of selenium are retained by the body to a greater degree than inorganic forms. There is some evidence that a homeostatic mechanism operates at physiologic levels of inorganic selenium intake to limit the amounts of selenium retained in the tissues.

Selenium is metabolized by a combination of reduction and methylation processes. Methylated metabolites of selenium include trimethyl selenonium ion, the major urinary metabolite of selenium, and dimethyl selenide, the volatile selenium metabolite produced under conditions of

Summary and Conclusions

selenium toxicity. Reduction of selenium *in vivo* is probably accomplished by its reaction with the sulfhydryl groups, either of proteins or of thiols of low molecular weight, such as glutathione, to form selenotrisulfide derivatives. The exact chemical nature of the "protein-bound" selenium in tissues is still not known with certainty. The evidence for the biosynthesis of seleno-amino acids from inorganic selenium by monogastric animals thus far appears rather tenuous.

Several important metabolic interrelationships exist between selenium and other elements of ecologic interest, such as mercury, cadmium, and arsenic. Under some conditions, the toxicity of selenium and these other elements is antagonistic; under other conditions, the toxicities are synergistic. These interactions provide a rich area of research for workers interested in mineral metabolism.

Several microorganisms are able to biosynthesize selenomethionine from inorganic selenium salts, but evidence for the formation of selenocystine is much less certain.

Although many microorganisms produce methylated selenium metabolites, these compounds are less toxic than the soluble inorganic salts of selenium. This is in contrast to the situation with mercury.

Selenium appears to be necessary for the functioning of certain bacterial enzymes, such as the formate dehydrogenase of *Escherichia coli*.

Some microorganisms seem to be able to adapt to high ambient concentrations of selenium, either by producing more enzyme to convert soluble selenium salts to the insoluble elemental selenium (selenoreductase) or by decreasing the uptake of selenium by altering the permeability of cell membranes.

NUTRITION

Selenium deficiency has been produced in rats, sheep, poultry, and lower primates, even when these animals are fed diets that are adequate in vitamin E. Rats must be depleted through two or three generations on diets containing 20 ppb of selenium in order for them to produce young that are deficient in the element. Deficiency symptoms noted in such animals include cataract, vascular hypoplasia, alopecia, and reproductive failure. In the chick, selenium deficiency is manifested by a severe pancreatic atrophy.

Selenium deficiency became a practical agricultural problem shortly after World War II because of changes in animal nutrition. To deal with this problem, the Food and Drug Administration has recently approved the use of selenium as a feed additive.

Although the importance of selenium-deficiency diseases in animals is well established, little is known concerning the role of selenium in human nutrition.

REPRODUCTION

Selenium deficiency or excess causes adverse effects on reproduction, which have usually been attributed to effects on the female. Further studies are needed to establish whether these detrimental effects are primarily on the reproductive system or are secondary to generalized emaciation or debilitation. Recent experiments implicate an essential role for selenium in the morphology and functioning of spermatozoa.

VASCULAR SYSTEM

A review of the effects of selenium deficiency and selenosis reveals a repetitious undercurrent of vascular-type lesions suggesting a primary vascular involvement in selenium disorders. Experimental evidence now available suggests that selenium deficiency causes endothelial hypoplasia and degeneration in tissues, accompanied by marginal vascular supply and oxygen dependence.

SELENOSIS

Selenium compounds were known by the mid-nineteenth century to be toxic, but it was not until about 1930 that the element was found to occur naturally in plants growing on soils of certain areas in amounts toxic to livestock.

The acute type of selenium poisoning results in the field from the consumption of highly seleniferous indicator plants and in experimental animals from the administration of high levels of selenium compounds or the feeding of highly seleniferous plant materials. Chronic selenium poisoning of the alkali disease type results in the field from the ingestion of feeds containing 5–40 ppm of the element or in the laboratory from the feeding or administration of similar levels of seleniferous feeds or selenium compounds. It has been suggested that chronic selenium poisoning of the blind staggers type is caused in the field by the consumption of subacutely toxic amounts of indicator plants, but whether selenium is the cause of this syndrome is questionable. Actually, the toxicity of the element generally increases gradually as its intake increases, so that differentiation of types of poisoning is sometimes difficult.

Summary and Conclusions

A number of factors alter the toxic effects of selenium on animals. These include the route of administration, the rate of intake or administration, the species and age of the animals, and the chemical form of the element. In addition, a number of criteria are used in measuring toxicity. These include macroscopic, microscopic, and biochemical observations. Thus, there are difficulties in attempting to establish at what level the intake or administration of selenium becomes harmful.

For instance, while as little as 0.25 ppm of selenium in the diet or 0.01 ppm in the water have been reported to cause physiologic or histologic changes in some experimental animals, it appears that the continuous intake of at least 1 ppm of dietary selenium will not normally have significant adverse physiologic effects.

At present, the only practical preventive measure for selenium poisoning in the field is avoidance of excessively seleniferous feeds. Likewise, no effective treatment of poisoned animals, other than removal of the source of selenium from the diet, is known.

SELENIUM POISONING IN MAN

Only a few reports of the poisoning of man through the consumption of seleniferous foods exist, and these fail to establish that selenium caused the signs of toxicity observed.

Other reports deal with industrial poisonings in which selenium in various forms is most commonly absorbed through the lungs or skin. In general, the forms of selenium involved have been the element itself, hydrogen selenide, selenium dioxide, and selenium oxychloride, and the signs of poisoning have been pallor, nervousness, coated tongue, depression, dermatitis, gastrointestinal disturbances, and garlic odor of the breath. No deaths among industrial workers have been reported to have resulted from selenium intoxication.

It has been suggested that low levels of dietary selenium cause dental caries. This matter deserves further study, but at this time it seems doubtful that selenium is significantly cariogenic.

CARCINOGENESIS

Six studies in the literature have examined the question of whether selenium is able to cause cancer. Three studies found that selenium is carcinogenic, and three found that it is not. However, a critical evaluation of these six trials showed significant experimental faults in the studies that claimed to find a carcinogenic role for selenium. Such faults were not present in the studies that found no carcinogenic activity for selenium. The scientific evidence available at this time suggests that selenium is not carcinogenic.

Some epidemiologic data and laboratory experiments indicate that selenium may have anticarcinogenic properties. An increased incidence of cancer in human beings has been correlated with a decreased level of selenium in the blood. Also, selenium has been found to have some inhibitory effect on the development of tumors in rodents given certain carcinogens. Further work along these lines is needed.

PHYSIOLOGIC ROLE

The only fully documented physiologic role for selenium in mammalian systems at present is in the enzyme glutathione peroxidase. However, recent studies suggest that selenium may also function in electron-transfer mechanisms and may involve nonheme iron proteins.

HUMAN MEDICAL USES

Selenium has several human medical uses. Selenium sulfide has been used as an antidandruff preparation and as an antifungal agent in tinea versicolor. The radionuclide form, ^{75}Se, usually as selenomethionine, is used for scanning of organs and tissues. Its primary use is for detection of tumor masses and assessment of placental competence.

Sampling and Analysis

A number of satisfactory methods, based on a variety of chemical or physical principles, are available for analyzing natural materials for selenium. The most commonly used procedures involve the use of wet digestion, reaction with 2,3-diaminonaphthalene, and fluorometric measurement of the extracted piazselenol. Neutron activation analysis and spark-source mass spectrometry require sophisticated equipment that is not available in most laboratories, but where multielement analysis of samples is required, they are very useful. Atomic-absorption spectrometry may become a sensitive and much-used method for selenium analysis. Any method selected should be tested in each laboratory on samples of known selenium content before it is used for routinely analyzing samples, and samples should be prepared in such a way as to avoid contamination and ensure representative analyses.

CONCLUSIONS

Although selenium is highly toxic in many of its chemical forms, a number of factors suggest that it probably is not a significant pollution problem. The bulk of the industrial uses of selenium are such that only small amounts of the element are injected into the ecosphere. Burning of coal and oil is estimated to account for nearly 70% of the selenium emitted into the atmosphere, but dispersion of selenium as a result of fossil fuel combustion does not appear to be an important pollution problem. The chemical forms of selenium liberated by the combustion are either insoluble (elemental selenium) or are bound tightly by soil colloids (selenium dioxide). Moreover, these likely forms of airborne selenium tend to aggregate in particulate form and therefore would not be expected to be as widely dispersed as such gaseous contaminants as sulfur dioxide.

The pesticidal uses of selenium in agriculture have a negligible impact on the environment. The projected use of selenium as an additive to animal feeds is considered to have little potential for contributing to the burden of this element in the environment. Pollution of waterways by selenium is likely to be maintained at a low level because of the precipitation of selenite selenium by the oxides of manganese and iron. Although the selenium content of foods varies widely, the variation is not extreme in either direction, and little concern regarding selenium deficiency or selenium toxicity in human beings is warranted. This statement, of course, does not apply to populations that may be subsisting entirely on foods grown in seleniferous regions. There is little evidence to indicate any biomagnification of selenium in the food chain.

These reassuring statements must not obscure the fact that in many areas we are ignorant concerning the ecologic impact of selenium. For example, little quantitative information exists concerning the natural cycling of selenium. Therefore, it is not possible to determine at this time whether selenium is becoming more or less available to man through the food supply as a result of enrichment or depletion of selenium in soils. Also, the nature and extent of industrial cycling of selenium are largely unknown, since selenium emissions are not generally a matter of record. It is difficult to assess the potential harm of airborne selenium, since practically nothing is known concerning the toxicity or metabolism of selenium compounds taken in via the lungs. Selenium has several profound metabolic interactions with other elements of ecologic concern, such as mercury, cadmium, and arsenic. Under some conditions, these interactions can be beneficial, but under other conditions they are distinctly harmful. Finally, although reliable analytical methods for selenium already exist, more convenient procedures will have to be developed before selenium determinations become part of any routine screening program. Undoubtedly, the paucity of data on the selenium content of various environmental samples is partly due to the difficulty of performing selenium analyses.

8

Recommendations

IMPROVED MONITORING OF SELENIUM IN THE ENVIRONMENT

RECOMMENDATION 1

More complete data are needed on the selenium content of fossil fuels and air and water samples. Unfortunately, the lack of a convenient method for selenium assay probably will impede the implementation of this recommendation (see Recommendation 3).

RECOMMENDATION 2

Additional information is needed on the natural and industrial cycling and industrial emissions of selenium.

Our relative ignorance concerning the ecologic fate of selenium prevents us from making absolute statements regarding its cycling in the ecosphere. Although the broad qualitative pathways of the natural cycling of selenium appear to be well outlined, we know much less about the quantitative aspects of such cycling. A similar state of affairs exists in the case of industrial cycling and industrial emissions. Once again, the analytical problems associated with selenium (Recommendation 3) have undoubtedly contributed directly to the lack of hard data concerning the environmental distribution of the element, and further progress in this field would be greatly facilitated by better analytic methodology.

RECOMMENDATION 3

Research to develop convenient, reliable methods for selenium analysis suitable for screening programs should be encouraged.

As Recommendations 1 and 2 indicate, there is a need to develop a convenient, accurate assay for selenium. Existing methods are either cumbersome and tedious (fluorometry) or require a sizable investment in instrumentation and facilities (neutron activation). Atomic-absorption spectrometry looks promising for the future, but preparation of samples will still present a problem.

BETTER KNOWLEDGE ABOUT SELENIUM TOXICOLOGY, METABOLISM, AND NUTRITION

RECOMMENDATION 4

The toxicology and metabolism of selenium compounds absorbed via the lungs should be investigated.

As in many other fields of toxicology, the toxicity of selenium has been studied mainly by injecting or feeding compounds of the element. Little work has been based on administering compounds via the pulmonary route. Judgments concerning the relative hazards of airborne selenium are extremely difficult to make in the virtual absence of fundamental data about the absorption and toxicity of selenium compounds that are inhaled.

RECOMMENDATION 5

Possible harmful effects of long-term, low-level exposure to selenium should be studied.

Most selenium toxicity experiments in the past have been either acute-injection studies or chronic-feeding studies. Toxicologists have paid little attention to the possibility that selenium may have detrimental effects at levels lower than those that cause death or depress growth. More sensitive criteria are needed for assessing selenium poisoning. Recent work has suggested that selenium at levels lower than those generally considered toxic can cause hepatic changes in rats. These results indicate the necessity for additional efforts to establish no-effect exposure levels of selenium.

Recommendations

RECOMMENDATION 6

Additional basic research is needed to elucidate the molecular mechanism of selenium toxicity.

Selenium has been shown to inhibit a number of biologic processes and enzyme systems, but there is little agreement among toxicologists as to which reaction constitutes the fundamental biochemical lesion in selenium toxicity. A better understanding of the primary target in selenium poisoning might lead to the development of a useful diagnostic indicator of selenium toxicity.

RECOMMENDATION 7

Intensive effort is required to clarify the metabolic interactions of selenium with other elements of ecologic concern, such as mercury, cadmium, and arsenic.

Under certain conditions, selenium can provide full protection against the toxicity of mercury and cadmium, and arsenic is able to diminish the severity of selenium poisoning. The biochemical basis of these metabolic antagonisms is poorly understood. Clarification of the molecular mechanisms behind them could conceivably lead to the discovery of new antidotes for certain types of heavy metal poisoning. It has been suggested that one of the "nutritional" roles of selenium may be to sequester traces of toxic heavy metals that occur in the environment. However, certain metabolites of selenium can have synergistic rather than antagonistic effects with mercury; hence, one must be careful in generalizing about the effects of selenium in the presence of heavy metals.

RECOMMENDATION 8

The possibility that selenium has anticarcinogenic effects should be explored further.

Evidence that selenium possesses anticarcinogenic properties is sufficient to warrant further investigation. Both laboratory and epidemiologic approaches should be taken.

RECOMMENDATION 9

The importance of selenium in human nutrition should be determined.

A great deal is known concerning the role of selenium in the nutrition of farm and laboratory animals, and the symptoms of selenium deficiency are clearly defined in these species. But relatively little is known about the role of selenium in human nutrition, and, because of the great diversity of deficiency diseases in animals, the nature of a hypothetic selenium deficiency in man is hard to predict. One could reasonably suppose that selenium is involved in such human medical problems as cancer, cataract, diseases of the liver, cardiovascular or muscular diseases, and the aging process.

RECOMMENDATION 10

Additional fundamental work on the physiologic role of selenium should be carried out.

Although characterization of selenium as a constituent of the active site of the enzyme glutathione peroxidase represents one of the towering achievements of selenium biochemistry, some workers feel that selenium may have other roles in metabolism. Research directed at this question should at least help explain the wide variety of symptoms observed in selenium-deficient animals, and it could contribute to our basic understanding of the human diseases mentioned in Recommendation 9.

References

1. Abu-Erreish, G. M. On the Nature of Some Selenium Losses from Soils and Waters. M.S. thesis, Brookings: South Dakota State University, 1967. 61 pp.
2. Abu-Erreish, G. M., E. I. Whitehead, and O. E. Olson. Evolution of volatile selenium compounds from soils. Soil Sci. 106:415–420, 1968.
3. Ageton, R. W. Selenium, pp. 713–721. In Mineral Facts and Problems. U.S. Bureau Of Mines Bulletin No. 650. Washington, D.C.: U.S. Department of the Interior, 1970.
4. Agnew, J. E., G. R. Youngs, and I. A. D. Bouchier. Studies of pancreatic morphology and function with ^{75}Se-selenomethionine. Proc. Brit. Inst. Radiol. 44:403, 1971.
5. Albert, S. N., C. A. Albert, E. F. Hirsch, I. N. Brecher, and P. Numerof. Selenate as a substitute for sulfate in the measurement of extracellular fluid volume. J. Nucl. Med. 7:290–293, 1966.
6. Albright, S. D., III, and J. M. Hitch. Rapid treatment of tinea versicolor with selenium sulfide. Arch. Derm. 93:460–462, 1966.
7. Allaway, W. H. Control of the environmental levels of selenium, pp. 181–206. In D. D. Hemphill, Ed. Trace Substances in Environmental Health—II. Proceedings of University of Missouri's 2nd Annual Conference on Trace Substances in Environmental Health. Columbia: University of Missouri, 1968.
8. Allaway, W. H. Selenium in the food chain. Cornell Vet. 63:151–170, 1973.
9. Allaway, W. H. Sulphur-selenium relationships in soils and plants. Sulphur Inst. J. 6(3):3–5, 1970.
10. Allaway, W. H., and E. E. Cary. Determination of submicrogram amounts of selenium in biological material. Anal. Chem. 36:1359–1362, 1964.
11. Allaway, W. H., E. E. Cary, and C. F. Ehlig. The cycling of low levels of selenium in soils, plants and animals, pp. 273–296. In O. H. Muth, Ed. Symposium: Selenium in Biomedicine. First International Symposium, Oregon State University, 1966. Westport, Conn.: AVI Publishing Co., 1967.
12. Allaway, W. H., J. Kubota, F. Losee, and M. Roth. Selenium, molybdenum, and vanadium in human blood. Arch. Environ. Health 16:342–348, 1968.
13. Allaway, W. H., D. P. Moore, I. E. Oldfield, and O. H. Muth. Movement of physi-

ological levels of selenium from soils through plants to animals. J. Nutr. 88:411–418, 1966.
14. Altschul, R. Endothelium. Its Development, Morphology, Function and Pathology. New York: The Macmillan Co., 1954. 171 pp.
15. American Conference of Governmental Industrial Hygienists. Documentation of the Threshold Limit Values for Substances in Workroom Air. (3rd ed.) Cincinnati, Ohio: American Conference of Governmental Industrial Hygienists, 1971. 286 pp.
16. Amor, A. J., and P. Pringle. A review of selenium as an industrial hazard. Bull. Hyg. 20:239–241, 1945.
17. Anderson, H. D., and A. L. Moxon. The excretion of selenium by rats on a seleniferous wheat ration. J. Nutr. 22:103–108, 1941.
18. Anderson, M. S., H. W. Lakin, K. C. Beeson, F. F. Smith, and E. Thacker. Selenium in Agriculture. U.S. Department of Agriculture Handbook No. 200. Washington, D.C.: U.S. Government Printing Office, 1961. 65 pp.
19. Anghileri, L. J., and R. Marqués. Fate of injected Se 75-methionine and Se 75-cystine in mice. Arch. Biochem. Biophys. 111:580–582, 1965.
20. Archer, V. E., and E. Luell. Effect of selenium sulfide suspension on hair roots. J. Invest. Derm. 35:65–67, 1960.
21. Argenbright, L. P. Smelter pollution control—Facts and problems. Mining Congress J. 57:24–28, 1971.
22. Arnold, R. L., O. E. Olson, and C. W. Carlson. Dietary selenium and arsenic additions and their effects on tissue and egg selenium. Poultry Sci. 52:847–854, 1973.
23. Arnold, R. L., O. E. Olson, and C. W. Carlson. Selenium withdrawal and egg selenium content. Poultry Sci. 51:341–342, 1972.
24. Asher, C. J., C. S. Evans, and C. M. Johnson. Collection and partial characterization of volatile selenium compounds from *Medicago sativa* L. Austral. J. Biol. Sci. 20:737–748, 1967.
25. Ashkar, F. S., J. L. Naya, and E. M. Smith. Parathyroid scanning with ^{75}Se-selenomethionine and glucagon stimulation. J. Nucl. Med. 12:751–753, 1971.
26. Atkins, H. L., W. Hauser, and J. F. Klopper. Effect of carrier on organ distribution of ^{75}Se-selenomethionine. Metabolism 20:1052–1056, 1971.
27. Awschalom, M. Datos sobre la influencia del selenio en al vegetación cuando sustituye el ion sulfúrico en el liquido nutritivo de Knop. Influencia del selenito sódico en la vida de los microorganismos. Rev. Fac. Agron. La Plata 14:122–162, 1921.
28. Awwad, H. K., E. J. Potchen, S. J. Adelstein, and J. B. Dealy, Jr. The regional distribution of Se^{75}-selenomethionine in the rat. Metab. Clin. Exp. 15:370–378, 1966.
29. Ayres, S., III, and S. Ayres, Jr. Selenium sulfide ointment in the treatment of seborrheic dermatitis of the glabrous skin. A.M.A. Arch. Derm. Syphil. 69:615–616, 1954.
30. Badiello, R., and A. Breccia. Radioprotection by selenium-containing compounds, pp. 103–107. In H. Moroson and M. Quintiliani, Eds. Radiation Protection and Sensitization. Proceedings of the International Symposium on Radiosensitizing and Radioprotective Drugs. 2nd, Rome, New York: Barnes and Noble, Inc , 1970.
31. Badiello, R., A. Trenta, M. Mattii, and S. Moretti. Azione radioprotettiva dei seleno-derivati: Effetto della selenourea "in vivo." Med. Nucl. Radiobiol. Lat. 10:57–68, 1967.
32. Baird, R. B., S. Pourian, and S. M. Gabrielian. Determination of trace amounts of selenium in wastewaters by carbon rod atomization. Anal. Chem. 44:1887–1889, 1972.

References

33. Bauer, F. K. Experimentelle Vergiftung mit Selen unter besonderer Berücksichtigung der Wirkung auf das Knochensystem. Arch. Gewerbepath. Gewerbehyg. 10:117–132, 1940.
34. Beath, O. A. The occurrence of selenium and seleniferous vegetation in Wyoming. Part II. Seleniferous vegetation of Wyoming, pages 29-64. In Wyoming Agricultural Experiment Station Bulletin No. 221: Laramie: University of Wyoming Agricultural Experiment Station, 1937.
35. Beath, O. A. The Story of Selenium in Wyoming. Laramie: University of Wyoming, not dated, probably 1962. 35 pp.
36. Beath, O. A. Toxic vegetation growing on the salt wash sandstone member of the Morrison formation. Amer. J. Botany 30:698–707, 1943.
37. Beath, O. A., J. H. Draize, and H. F. Eppson. Three poisonous vetches, pp. 1–23. In Wyoming Agricultural Experiment Station Bulletin No. 189. Laramie: University of Wyoming Agricultural Experiment Station, 1932.
38. Beath, O. A., J. H. Draize, H. F. Eppson, C. S. Gilbert, and O. C. McCreary. Certain poisonous plants of Wyoming activated by selenium and their association with respect to soil types. J. Amer. Pharm. Assoc. 23:94–97, 1934.
39. Beath, O. A., and H. F. Eppson. The form of selenium in some vegetation, pp. 1–17. In Wyoming Agricultural Experiment Station Bulletin No. 278. Laramie: University of Wyoming Agricultural Experiment Station, 1947.
40. Beath, O. A., H. F. Eppson, and C. S. Gilbert. Selenium and Other Toxic Minerals in Soils and Vegetation. Wyoming Agricultural Experiment Station Bulletin No. 206. Laramie: University of Wyoming Agricultural Experiment Station, 1935. 55 pp.
41. Beath, O. A., H. F. Eppson, and C. S. Gilbert. Selenium distribution in and seasonal variation of type vegetation occurring on seleniferous soils. J. Amer. Pharm. Assoc. 26:394–405, 1937.
42. Beath, O. A., C. S. Gilbert, and H. F. Eppson. The use of indicator plants in locating seleniferous areas of western United States. I. General. Amer. J. Botany 26:257–269, 1939.
43. Beath, O. A., A. F. Hagner, and C. S. Gilbert. Some Rocks of High Selenium Content. Wyoming Geological Survey Bulletin No. 36. Laramie: The Geological Survey of Wyoming, University of Wyoming, 1946. 23 pp.
44. Bera, B. C., and M. M. Chakrabartty. Spectrophotometric determination of selenium with 2-mercaptobenzothiazole. Analyst 93:50–55, 1968.
45. Berenshtein, F. Ya., and S. F. Aleshko. Effect of selenium on lipid metabolism. Khim. Sel. Khoz. 6:937–940, 1968. (in Russian)
46. Bereston, E. S. Use of selenium sulfide shampoo in seborrheic dermatitis. J.A.M.A. 156:1246–1247, 1954.
47. Bertine, K. K., and E. D. Goldberg. Fossil fuel combustion and the major sedimentary cycle. Science 173:233–235, 1971.
48. Betteridge, D. Determination of Selenium in Hair by Neutron-Activation Analysis. United Kingdom Atomic Energy Research Establishment Report No. AERE-R 4881. London, England: Her Majesty's Stationery Office, 1965. 15 pp.
49. Bidstrup, P. L. Toxicity of Mercury and Its Compounds. Amsterdam: Elsevier Publishing Co., 1964, 112 pp.
50. Bieri, J. G., and C. J. Pollard. Serum protein changes in vitamin E-deficient chicks. J. Nutr. 69:301–305, 1959.
51. Bioassay of Selenium Compounds for Carcinogenesis in Rats. Final Report. Departments of Agricultural Chemistry and Veterinary Medicine. Corvallis: Oregon State University, 1966. 131 pp.

52. Bisbjerg, B. Studies on Selenium in Plants and Soils. Risø Report No. 200. Roskilde: Danish Atomic Energy Commission, Research Establishment Risø, 1972, 150 pp.
53. Bisbjerg, B., and G. Gissel-Nielsen. The uptake of applied selenium by agricultural plants. 1. The influence of soil type and plant species. Plant Soil 31:287-298, 1969.
54. Bisbjerg, B., P. Jochumsen, and N. O. Rasbech. Selenium content in organs, milk, and fodder of the cow. Nord. Veterinaermed. 22:532-535, 1970.
55. Black, R. J., A. J. Muhich, A. J. Klee, H. L. Hickman, Jr., and R. D. Vaughan. The National Solid Wastes Survey. An Interim Report. Presented at the 1968 Annual Meeting of the Institute for Solid Wastes of the American Public Works Association, Miami Beach, Florida, October 24, 1968. Cincinnati, Ohio: U.S. Department of Health, Education, and Welfare, Bureau of Solid Waste Management, 1968. 53 pp.
56. Blau, M. Biosynthesis of [^{75}Se] selenomethionine and [^{75}Se] selenocystine. Biochim. Biophys. Acta 49:389-390, 1961.
57. Blau, M., and M. A. Bender. Se 75-selenomethionine for visualization of the pancreas by isotope scanning. Radiology 78:974, 1962.
58. Blau, M., R. F. Maske, and M. A. Bender. Clinical experience with Se^{75} selenomethionine for pancreas visualization. J. Nucl. Med. 3:202, 1962.
59. Blaxter, K. L. The effect of selenium on lamb growth: Cooperative experiments on Scottish farms. Proc. Nutr. Soc. 21:xix, 1962.
60. Blincoe, C. Whole-body turnover of selenium in the rat. Nature 186:398, 1960.
61. Bonhorst, C. W. Anion antagonisms in yeast as indicators of the mechanism of selenium toxicity. J. Agric. Food Chem. 3:700-703, 1955.
62. Bonhorst, C. W., and J. J. Mattice. Colorimetric determination of selenium in biological materials. Anal. Chem. 31:2106-2107, 1959.
63. Bonucci, E., and R. Sadun. Experimental calcification of the myocardium: Ultrastructural and histochemical investigations. Amer. J. Path. 71:167-192, 1973.
64. Bowen, H. J. M., and P. A. Cawse. The determination of selenium in biological material by radioactivation. Analyst 88:721-726, 1963.
65. Bowen, W. H. The effects of selenium and vanadium on caries activity in monkeys (*M. irus*). J. Irish Dent. Assoc. 18:83-89, 1972.
66. Braun, H. A., L. M. Lusky, and H. O. Calvery. The efficacy of 2,3-dimercaptopropanol (BAL) in the therapy of poisoning by compounds of antimony, bismuth, chromium, mercury and nickel. J. Pharmacol. Exp. Ther. 87(Suppl.):119-125, 1946.
67. Brodsky, I., E. M. Ross, S. B. Kahn, and G. Perkov. Platelet and fibrinogen kinetics with ^{75}Se selenomethionine in thrombocytopenic states. Brit. J. Haematol. 22:589-598, 1972.
68. Brown, D. G., and R. F. Burk. Selenium retention in tissues and sperm of rats fed a torula yeast diet. Fed. Proc. 31:692, 1972.
69. Brown, D. G., and R. F. Burk. Selenium retention in tissues and sperm of rats fed a torula yeast diet. J. Nutr. 103:102-108, 1973.
70. Brown, D. G., R. F. Burk, R. J. Seely, and K. W. Kiker. Effect of dietary selenium on the gastrointestinal absorption of ^{75}SeO$_3$ in the rat. Int. J. Vit. Nutr. 42:588-591, 1972.
71. Brown, J. H., and S. H. Pollock. Stabilization of hepatic lysosomes of rats by vitamin E and selenium *in vivo* as indicated by thermal labilization of isolated lysosomes. J. Nutr. 102:1413-1419, 1972.
72. Brown, M. J., and D. L. Carter. Leaching of added selenium from alkaline soils as influenced by sulfate. Soil Sci. Soc. Amer. Proc. 33:563-565, 1969.
73. Brown, P. W., W. Sircus, A. N. Smith, A. A. Donaldson, I. W. Dymock, C. W. A.

References

Falconer, and W. P. Small. Scintillography in diagnosis of pancreatic disease. Lancet 1:160–163, 1968.
74. Broyer, T. C., C. M. Johnson, and R. P. Huston. Selenium and nutrition of *Astragalus*. I. Effects of selenite or selenate supply on growth and selenium content. Plant Soil 36:635–649, 1972.
75. Broyer, T. C., D. C. Lee, and C. J. Asher. Selenium nutrition of green plants. Effect of selenite supply on growth and selenium content of alfalfa and subterranean clover. Plant Physiol. 41:1425–1428, 1966.
76. Brusa, A., and G. P. Oneto. Sur les altérations de la structure de l'ovaire et des cellules-oeuf déterminées par le sélénium. C. R. Assoc. Anat. 41:333–349, 1954.
77. Buchan, R. F. Industrial selenosis. A review of the literature, report of five cases and a general bibliography. Occup. Med. 3:439–456, 1947.
78. Buchanan-Smith, J. G., E. C. Nelson, B. I. Osburn, M. E. Wells, and A. D. Tillman. Effects of vitamin E and selenium deficiencies in sheep fed a purified diet during growth and reproduction. J. Animal Sci. 29:808–815, 1969.
79. Buchanan-Smith, J. G., E. C. Nelson, and A. D. Tillman. Effect of vitamin E and selenium deficiencies on lysosomal and cytoplasmic enzymes in sheep tissues. J. Nutr. 99:387–394, 1969.
80. Buchanan-Smith, J. G., B. A. Sharp, and A. D. Tillman. Tissue selenium concentrations in sheep fed a purified diet. Can. J. Physiol. Pharmacol. 49:619–621, 1971.
81. Burk, R., Jr., W. N. Pearson, R. P. Wood II, and F. Viteri. Blood selenium levels and in vitro red blood cell uptake of ^{75}Se in kwashiorkor. Amer. J. Clin. Nutr. 20:723–733, 1967.
82. Burk, R. F., D. G. Brown, R. J. Seely, and C. C. Scaief III. Influence of dietary and injected selenium on whole-body retention, route of excretion, and tissue retention of ^{75}SeO$_3^{2-}$ in the rat. J. Nutr. 102:1049–1055, 1972.
83. Burk, R. F., R. J. Seeley, and K. W. Kiker. Selenium: Dietary threshold for urinary excretion in the rat. Proc. Soc. Exp. Biol. Med. 142:214–216, 1973.
84. Burk, R. F., Jr., R. Whitney, H. Frank, and W. N. Pearson. Tissue selenium levels during development of dietary liver necrosis in rats fed torula yeast diets. J. Nutr. 95:420–428, 1968.
85. Burkhanov, A. I., and R. K. Zhakenova. Changes in the lungs during intratracheal administration of elemental selenium to rats. Zdravookhr. Kaz. 33(3):52–53, 1973. (in Russian)
86. Burriel-Martí, F., F. Lucena-Conde, and S. Arribas-Jimeno. Mercurious salts as reductimetric reagents for titrations in alkaline solution. I. Titration of ferricyanide. Anal. Chim. Acta 10:301–309, 1954.
87. Burton, V., R. F. Keeler, K. F. Swingle, and S. Young. Nutritional muscular dystrophy in lambs—Selenium analysis of maternal, fetal, and juvenile tissues. Amer. J. Vet. Res. 23:962–965, 1962.
88. Butler, G. W., and P. J. Peterson. Aspects of the faecal excretion of selenium by sheep. N.Z. J. Agric. Res. 4:484–491, 1962.
89. Butler, G. W., and P. J. Peterson. Uptake and metabolism of inorganic forms of selenium-75 by *Spiradela oligorrhiza*. Austral. J. Biol. Sci. 20:77–86, 1967.
90. Büttner, W. Effects of some trace elements on fluoride retention and dental caries. Arch. Oral. Biol. 6 (Special Suppl.):40–49, 1961.
91. Byard, J. L. The Metabolism of Sodium Selenite in the Rat. Ph.D. thesis, Madison: University of Wisconsin, 1968. 101 pp.
92. Byard, J. L. Trimethyl selenide. A urinary metabolite of selenite. Arch. Biochem. Biophys. 130:556–560, 1969.

93. Byers, H. G. Selenium Occurrence in Certain Soils in the United States with a Discussion of Related Topics. U.S. Department of Agriculture Technical Bulletin No. 482. Washington, D.C.: U.S. Department of Agriculture, 1935. 47 pp.
94. Byers, H. G. Selenium Occurrence in Certain Soils in the United States with a Discussion of Related Topics. Second Report. U.S. Department of Agriculture Technical Bulletin No. 530. Washington, D.C.: U.S. Department of Agriculture, 1936. 78 pp.
95. Byers, H. G., and H. W. Lakin. Selenium in Canada. Can. J. Res. 17:364–369, 1939.
96. Byers, H. G., J. T. Miller, K. T. Williams, and H. W. Lakin. Selenium Occurrence in Certain Soils in the United States with a Discussion of Related Topics. Third Report. U.S. Department of Agriculture Technical Bulletin No. 601. Washington, D.C.: U.S. Department of Agriculture, 1938. 74 pp.
97. Byers, H. G., K. T. Williams, and H. W. Lakin. Selenium in Hawaii and its probable source in the United States. Ind. Eng. Chem. Ind. Ed. 28:821–823, 1936.
98. Cadell, P. B., and F. B. Cousins. Urinary selenium and dental caries. Nature 185:863–864, 1960.
99. Calvert, C. C., M. C. Nesheim, and M. L. Scott. Effectiveness of selenium in prevention of nutritional muscular-dystrophy in the chick. Proc. Soc. Exp. Biol. Med. 109:16–18, 1962.
100. Cameron, C. A. Preliminary note on the absorption of selenium by plants. Roy. Dublin Soc. Sci. Proc. 2:231–233, 1880.
101. Cantor, A. H., M. L. Scott, and T. Noguchi. Biological availability of selenium in feedstuffs and selenium compounds for prevention of exudative diathesis in chicks. J. Nutr. 105:96–105, 1975.
102. Caravaggi, C., and F. L. Clark. Mortality in lambs following intramuscular injection of sodium selenite. Austral. Vet. J. 45:383, 1969.
103. Caravaggi, C., F. L. Clark, and A. R. B. Jackson. Acute selenium toxicity in lambs following intramuscular injection of sodium selenite. Res. Vet. Sci. 11:146–149, 1970.
104. Carlson, C. W., P. L. Guss, and O. E. Olson. Selenium content of chick tissues as affected by arsenic. Poultry Sci. 41:1987–1989, 1962.
105. Carlson, C. W., W. Kohlmeyer, and A. L. Moxon. Arsenic fails to control selenium poisoning in turkeys. S. Dakota Farm Home Res. 3:20–22, 1951.
106. Caroli, J., G. Milhaud, J. E. de Saint Laurent, P. Hadchouel, and G. Theodoropoulos. Méthode originale de diagnostic des cancers du foie. La scintigraphie à la sélénométhionine. Ann. Med. Interne 122:747–754, 1971.
107. Carter, R. F. Acute selenium poisoning. Med. J. Austral. 1:525–528, 1966.
108. Cary, E. E., and W. H. Allaway. Selenium content of field crops grown on selenite-treated soils. Agron. J. 65:922–925, 1973.
109. Cary, E. E., and W. H. Allaway. The stability of different forms of selenium added to low-selenium soils. Soil Sci. Amer. Soc. Proc. 33:571–574, 1969.
110. Cary, E. E., G. A. Wieczorek, and W. H. Allaway. Reactions of selenite-selenium added to soils that produce low-selenium forages. Soil Sci. Soc. Amer. Proc. 31:21–26, 1967.
111. Caygill, C. P. J., J. A. Lucy, and A. T. Diplock. The effect of vitamin E on the intracellular distribution of the different oxidation states of selenium in rat liver. Biochem. J. 125:407–416, 1971.
112. Cerwenka, E. A., Jr., and W. C. Cooper. Toxicology of selenium and tellurium and their compounds. Arch. Environ. Health 3:189–200, 1961.
113. Challenger, F. Biological methylation. Adv. Enzymol. 12:429–491, 1951.
114. Chan, F. L. 4,5-Diamino-6-thiopyrimidine as an analytical reagent. I. Spectrophotometric determination of selenium. Talanta 11:1019–1029, 1964.
115. Chen, R. W., P. A. Wagner, W. G. Hoekstra, and H. E. Ganther. Affinity labelling

studies with ^{109}cadmium in cadmium-induced testicular injury in rats. J. Reprod. Fertil. 38:293-306, 1974.
116. Cherkes, L. A., S. G. Aptekar, and M. N. Volgarev. Tumors of the liver produced by selenium. Byul. Eksp. Biol. Med. 53(3):78-83, 1962. (in Russian)
117. Chernick, S. S., J. G. Moe, G. P. Rodnan, and K. Schwarz. A metabolic lesion in dietary necrotic liver degeneration. J. Biol. Chem. 217:829-843, 1955.
118. Chiquoine, A. D. Effect of cadmium chloride on the pregnant albino mouse. J. Reprod. Fertil. 10:263-265, 1965.
119. Chow, C. M., S. N. Nigam, and W. B. McConnell. Biosynthesis of Se-methylselenocysteine and S-methylcysteine in *Astragalus bisulcatus:* Effect of selenium and sulphur concentrations in the growth medium. Phytochemistry 10:2693-2698, 1971.
120. Christian, G. D., E. C. Knoblock, and W. C. Purdy. Polarographic determination of selenium in biological materials. J. Assoc. Agric. Chemists 48:877-884, 1965.
121. Chu, S. H., and D. D. Davidson. Potential antitumor agents. α- and β-2'-Deoxy-6-selenoguanosine and related compounds. J. Med. Chem. 15:1088-1089, 1972.
122. Claycomb, C. K., F. M. Sorenson, D. C. Gatewood, E. B. Jump, and M. E. Weaver. Further studies on the presence of Se^{75} in rat saliva and teeth after intracardiac injection of radioactive sodium selenite. J. Dent. Res. 40:504-510, 1961.
123. Clayton, C. C., and C. A. Baumann. Diet and azo dye tumors; effect of diet during a period when the dye is not fed. Cancer Res. 9:575-582, 1949.
124. Clinton, M., Jr. Selenium fume exposure. J. Ind. Hyg. Toxicol. 29:225-226, 1947.
125. Coch, E. H., and R. C. Greene. The utilization of selenomethionine by *Escherichia coli.* Biochim. Biophys. Acta 230:223-226, 1971.
126. Coleman, R. G., and M. Delevaux. Occurrence of selenium in sulfides from some sedimentary rocks of the western United States. Econ. Geol. 52:499-527, 1957.
127. Colton, M. W. Selenium-tocopherol treatment of chronic lameness in dogs. Mod. Vet. Pract. 46(10):92, 1965.
128. Cooper, W. C. Selenium toxicity in man, pp. 185-199. In O. H. Muth, Ed. Symposium: Selenium in Biomedicine. First International Symposium, Oregon State University, 1966. Westport, Conn.: AVI Publishing Co., 1967.
129. Cousen, A. The estimation of selenium in glass. J. Soc. Glass Technol. 7:303-309, 1923.
130. Cousins, F. B. Fluorimetric microdetermination of selenium in biological materials. Austral. J. Exp. Biol. Med. Sci. 38:11-16, 1960.
131. Cousins, F. B., and I. M. Cairney. Some aspects of selenium metabolism in sheep. Austral. J. Agric. Res. 12:927-942, 1961.
132. Cowie, D. B., and G. N. Cohen. Biosynthesis by *Escherichia coli* of active altered proteins containing selenium instead of sulfur. Biochim. Biophys. Acta 26:252-261, 1957.
133. Creech, B. C., G. L. Feldman, T. M. Ferguson, B. L. Reid, and J. R. Couch. Exudative diathesis and vitamin E deficiency in turkey poults. J. Nutr. 62:83-96, 1957.
134. Cresser, M. S., and T. S. West. Studies in the analytical chemistry of selenium: Absorptiometric determination with 2-mercaptobenzoic acid. Analyst 93:595-600, 1968.
135. Crystal, R. G. Elemental selenium: Structure and properties, pp. 13-27. In D. L. Klayman and W. H. H. Günther, Eds. Organic Selenium Compounds: Their Chemistry and Biology. New York: Wiley-Interscience, 1973.
136. Cukor, P., J. Walzcyk, and P. F. Lott. The application of isotope dilution analysis to the fluorimetric determination of selenium in plant material. Anal. Chim. Acta 30:473-482, 1964.

137. Cummins, L. M., and E. T. Kimura. Safety evaluation of selenium sulfide antidandruff shampoos. Toxicol. Appl. Pharmacol. 20:89–90, 1971.
138. Cummins, L. M., and J. L. Martin. Are selenocystine and selenomethionine synthesized *in vivo* from sodium selenite in mammals? Biochemistry 6:3162–3168, 1967.
139. Cummins, L. M., J. L. Martin, and D. D. Maag. An improved method for determination of selenium in biological material. Anal. Chem. 37:430–431, 1965.
140. Dam, H., and J. Glavind. Vitamin E und kapillarpermeabilitat. Naturwissenschaften 28:207, 1940.
141. Dams, R., J. A. Robbins, K. A. Rahn, and J. W. Winchester. Nondestructive neutron activation analysis of air pollution particulates. Anal. Chem. 42:861–867, 1970.
142. D'Angio, G. J., M. Loken, and M. Nesbit. Radionuclear (^{75}Se) identification of tumor in children with neuroblastoma. Radiology 93:615–617, 1969.
143. Danielli, J. F., M. Danielli, J. B. Fraser, P. D. Mitchell, L. N. Owen, and G. Shaw. Bal-Intrav: A new nontoxic thiol for intravenous injection in arsenical poisoning. Biochem. J. 41:325–333, 1947.
144. Davidson, D. F. Selenium in Some Epithermal Deposits of Antimony, Mercury, and Silver and Gold. U.S. Geological Survey Bulletin No. 1112-A. Washington, D.C.: U.S. Government Printing Office, 1960. 15 pp.
145. Davidson, D. F., and H. A. Powers. Selenium content of some volcanic rocks from western United States and Hawaiian Islands, pp. 69–81. In Contributions to Geochemistry 1958. U.S. Geological Survey Bulletin No. 1084. Washington, D.C.: U.S. Government Printing Office, 1961.
146. Davies, E. B., and J. H. Watkinson. Uptake of native and applied selenium by pasture species. I. Uptake of Se by browntop, ryegrass, cocksfoot, and white clover from Atiamuri sand. N.Z. J. Agric. Res. 9:317–327, 1966.
147. Davies, E. B., and J. H. Watkinson. Uptake of native and applied selenium by pasture species. II. Effects of sulfate and of soil type on uptake by clover. N.Z. J. Agric. Res. 9:641–652, 1966.
148. Davies, H. L. The effect of selenium and vitamin E on reproduction in Merino sheep in south-western Australia. J. Austral. Inst. Agric. Sci. 32:216–217, 1966.
149. Davis, W. E., and Associates. National Inventory of Sources and Emissions. Barium, Boron, Copper, Selenium and Zinc. Selenium. Section IV. Environmental Protection Agency. Office of Air Programs. Contract No. 68-02-0100. Leawood, Kans.: W. E. Davis and Associates, 1972. 50 pp.
150. Demeryere, D., and J. Hoste. *o*-Diamines as reagents for selenium. Part 1. 4-Dimethylamino-1,2-phenylenediamine. Anal. Chim. Acta 27:288–294, 1962.
151. De Salas, S. M. EL selenio en las aguas. 2a. Parte. Método para su determinación. Rev. Obras Sanit. Nación 11(119):264–275, 1947.
152. Deshmukh, G. S., and M. G. Bapat. Oxidation of selenite by alkaline ferricyanide with osmium tetroxide as a catalyst. Z. Anal. Chem. 156:273–276, 1957.
153. DeWitt, W. B., and K. Schwarz. Multiple dietary necrotic degeneration in the mouse. Experientia 14:28–30, 1958.
154. DiGiulio, W., and W. Beirwaltes. Parathyroid scanning with selenium^{75} labeled methionine. J. Nucl. Med. 5:417–427, 1964.
155. Dinkel, C. A., J. A. Minyard, and D. E. Ray. Effects of season of breeding on reproductive and weaning performance of beef cattle grazing seleniferous range. J. Animal Sci. 22:1043–1045, 1963.
156. Dinkel, C. A., J. A. Minyard, E. I. Whitehead, and O. E. Olson. Agricultural Research at the Reed Ranch Field Station. South Dakota Agricultural Experiment Station Circular No. 135. Brookings: South Dakota State College of Agriculture and Mechanic Arts Agricultural Experiment Station, 1957. 35 pp.

157. Diplock, A. T., H. Baum, and J. A. Lucy. The effect of vitamin E on the oxidation state of selenium in rat liver. Biochem. J. 123:721–729, 1971.
158. Diplock, A. T., and J. A. Lucy. The biochemical modes of action of vitamin E and selenium: A hypothesis. Fed. Eur. Biochem. Soc. Lett. 29:205–210, 1973.
159. Douglas, C. P. Assessment of placental competence. Scott. Med. J. 14:162–170, 1969.
160. Draize, J. H., and O. A. Beath. Observations on the pathology of "blind staggers" and "alkali disease." J. Amer. Vet. Med. Assoc. 86:753–763, 1935.
161. Dransfield, P. B., and F. Challenger. Studies on biological methylation. Part XV. The formation of dimethyl selenide in mould cultures in presence of D- and L-methionine, or of thetins, all containing the $^{14}CH_3$ group. J. Chem. Soc. 2:1153–1160, 1955.
162. Dubois, K. P., A. L. Moxon, and O. E. Olson. Further studies on the effectiveness of arsenic in preventing selenium poisoning. J. Nutr. 19:477–482, 1940.
163. Dudley, H. C. Selenium as a potential industrial hazard. U.S. Public Health Rep. 53:281–292, 1938.
164. Dudley, H. C. Toxicology of selenium. V. Toxic and vesicant properties of selenium oxychloride. U.S. Public Health Rep. 53:94–98, 1938.
165. Dudley, H. C., and J. W. Miller. Toxicology of selenium. IV. Effects of exposure to hydrogen selenide. U.S. Public Health Rep. 52:1217–1231, 1937.
166. Dudley, H. C., and J. W. Miller. Toxicology of selenium. VI. Effect of subacute exposure to hydrogen selenide. J. Ind. Hyg. Toxicol. 23:470–477, 1941.
167. Duhamel, B. C. Lésions histologiques dans l'intoxication par le sélénium colloidal et de l'acide sélénieux. C. R. Soc. Biol. 42:742–744, 1913.
168. Dumery, P.-R. Le sélénium, agent antimycosique. Presse Med. 61:1646, 1953.
169. Dutkiewicz, T., B. Dutkiewicz, and I. Balcerska. Dynamics of organ and tissue distribution of selenium after intragastric and dermal administration of sodium selenite. Bromatol. Chem. Toksykol. 4:475–481, 1972. (in Polish)
170. Dye, W. B., E. Bretthauer, H. J. Seim, and C. Blincoe. Fluorometric determination of selenium in plants and animals with 3,3'-diaminobenzidine. Anal. Chem. 35:1687–1693, 1963.
171. Eaton, R. P., and D. M. Kipnis. Incorporation of Se^{75} selenomethionine into a protein component of plasma very-low-density lipoprotein in man. Diabetes 21:744–753, 1972.
172. Eaton, S. B., D. J. Fleischli, J. J. Pollard, R. A. Nebesar, and M. S. Potsaid. Comparison of current radiologic approaches to diagnosis of pancreatic disease. New Engl. J. Med. 279:389–396, 1968.
173. Eaton, S. B., A. E. James, M. S. Potsaid, and G. L. Nardi. Evaluation of corrective surgery for pancreatitis by Se^{75} selenomethionine pancreatic imaging. Amer. J. Roentgenol. Radium Ther. Nucl. Med. 112:678–681, 1971.
174. Eddleston, A. L. W. F., M. O. Rake, A. P. Pagaltsos, S. B. Osburn, and R. Williams. ^{75}Se-selenomethionine in the scintiscan diagnosis of primary hepatocellular carcinoma. Gut 12:245–249, 1971.
175. Edgington, G., and H. G. Byers. Geology and biology of North Atlantic deep-sea cores between Newfoundland and Ireland, pp. 151–155. In Part 9. Selenium Content and Chemical Analyses. U.S. Geological Survey Professional Paper No. 196. Washington, D.C.: U.S. Government Printing Office, 1942.
176. Edmond, C. R. Dental caries etiology in New Guinea: Some contributions from analytical chemistry. Austral. Miner. Develop. Lab. Bull. 4:17–36, 1967.
177. Eggert, R. G., E. Patterson, W. T. Akers, and E. L. R. Stokstad. The role of vitamin E and selenium in the nutrition of the pig. J. Animal Sci. 16:1037, 1957.
178. Ehlig, C. F., W. H. Allaway, E. E. Cary, and J. Kubota. Differences among plant

species in selenium accumulation from soils low in available selenium. Agron. J. 60:43–47, 1968.
179. Eisenberg, B. C. Contact dermatitis from selenium sulfide shampoo. A.M.A. Arch. Derm. Syphil. 72:71–72, 1955.
180. Ellis, M. M., H. L. Motley, M. D. Ellis, and R. O. Jones. Selenium poisoning in fishes. Proc. Soc. Exp. Biol. Med. 36:519–522, 1937.
181. English, J. A. Experimental effects of thiouracil and selenium on the teeth and jaws of dogs. J. Dent. Res. 28:172–194, 1949.
182. Enoch, H. G., and R. L. Lester. Effects of molybdate, tungstate, and selenium compounds on formate dehydrogenase and other enzyme systems in *Escherichia coli*. J. Bacteriol. 110:1032–1040, 1972.
183. Ermakov, V. V. Selenium distribution in human organs and tissues. Byull. Eksp. Biol. Med. 59(3):61–62, 1965. (in Russian)
184. Euler, H. v., H. Hasselquist, and B. v. Euler. Biologisch wirksame Oligo-Elemente in organischer Bindung. Ark. Kemi 9:583–590, 1956.
185. Evans, C. S., C. J. Asher, and C. M. Johnson. Isolation of dimethyl diselenide and other volatile selenium compounds from *Astragalus racemosus* (Pursh.). Austral. J. Biol. Sci. 21:13–20, 1968.
186. Ewan, R. C., C. A. Baumann, and A. L. Pope. Effects of selenium and vitamin E on nutritional muscular dystrophy in lambs. J. Animal Sci. 27:751–756, 1968.
187. Ewan, R. C., C. A. Baumann, and A. L. Pope. Retention of selenium by growing lambs. J. Agric. Food Chem. 16:216–219, 1968.
188. Ewan, R. C., A. L. Pope, and C. A. Baumann. Elimination of fixed selenium by the rat. J. Nutr. 91:547–554, 1967.
189. Ewan, R. C., and M. E. Wastell. Effect of vitamin E and selenium on blood composition of the young pig. J. Animal Sci. 31:343–350, 1970.
190. Ewan, R. C., and M. E. Wastell, E. J. Bicknell, and V. C. Speer. Performance and deficiency symptoms of young pigs fed diets low in vitamin E and selenium. J. Animal Sci. 29:912–915, 1969.
191. Faulkner, A. G., C. E. Knoblock, and W. C. Purdy. The polarigraphic determination of selenium in urine. Clin. Chem. 7:22–29, 1961.
192. Fels, I. G., and V. H. Cheldelin. Selenate inhibition studies. II. The reversal of selenate inhibition in *E. coli*. Arch. Biochem. Biophys. 22:323–324, 1949.
193. Fels, I. G., and V. H. Cheldelin. Selenate inhibition studies. III. The role of sulfate in selenate toxicity in yeast. Arch. Biochem. Biophys. 22:402–405, 1949.
194. Fels, I. G., and V. H. Cheldelin. Selenate inhibition studies. IV. Biochemical basis of selenate toxicity in yeast. J. Biol. Chem. 185:803–811, 1950.
195. Ferretti, R. J., and O. A. Levander. Selenium content of grains and cereal products. J. Agric. Food Chem. 22:1049–1051, 1975.
196. Ferretti, R. J., and O. A. Levander. Selenium content of grains and cereal products as influenced by processing and milling. Fed. Proc. 33:693, 1974.
197. Ferrucci, J. T., R. A. Berke, and M. S. Potsaid. Se^{75} selenomethionine isotope lymphography in lymphoma: Correlation with lymphangiography. Amer. J. Roentgenol. Radium Ther. Nucl. Med. 109:793–802, 1970.
198. Filatova, V. S. Toxicity of selenic anhydride. Gig. Sanit. 5:18–23, 1951. (in Russian)
199. Filby, R. H., and W. L. Yakely. Determination of selenium in eye-lenses by neutron activation analysis using selenium-77m. Radiochem. Radioanal. Lett. 2:307–311, 1969.
200. Fimiani, R. Glutationemia nell'intossicazione cronica sperimentale da selenio. Folia Med. (Naples) 34:260–263, 1951.

References

201. Fink, H. Selenium content of skim milk powders and their inclination for producing dietetic liver necrosis, p. 19. In Abstracts of the Fifth International Congress on Nutrition, Washington, D.C., September, 1960.
202. Fink, H., I. Schlie, and K. Schwarz. Über die ernährungsbedingte Leberkrose und den Selengehalt von Milchpulvern verschiedener Herkunft. Z. Naturforsch. 22B: 1144–1149, 1968.
203. Fleming, G. A. Selenium in Irish soils and plants. Soil Sci. 94:28–35, 1962.
204. Fleming, G. A., and T. Walsh. Selenium occurrence in certain Irish soils and its toxic effects on animals. Proc. Roy. Irish Acad. 58(Sect. B):152–166, 1957.
205. Fleming, R. W., and M. Alexander. Dimethylselenide and dimethyltelluride formation by a strain of *Penicillium*. Appl. Microbiol. 24:424–429, 1972.
206. Flohe, L., W. A. Günzler, and H. H. Schock. Glutathione peroxidase: A selenoenzyme. Fed. Eur. Biochem. Soc. Lett. 32:132–134, 1973.
207. Franke, K. W. A new toxicant occurring naturally in certain samples of plant foodstuffs. I. Results obtained in preliminary feeding trials. J. Nutr. 8:597–608, 1934.
208. Franke, K. W. A new toxicant occurring naturally in certain samples of plant foodstuffs. II. The occurrence of the toxicant in the protein fraction. J. Nutr. 8:609–613, 1934.
209. Franke, K. W., R. Burris, and R. S. Hutton. A new colorimetric procedure adapted to selenium determination. Ind. Eng. Chem. Anal. Ed. 8:435, 1936.
210. Franke, K. W., and A. L. Moxon. A comparison of the minimum fatal doses of selenium, tellurium, arsenic and vanadium. J. Pharmacol. Exp. Ther. 58:454–459, 1936.
211. Franke, K. W., A. L. Moxon, W. E. Poley, and W. C. Tully. A new toxicant occurring naturally in certain samples of plant foodstuffs. XII. Monstrosities produced by the injection of selenium salts into hen's eggs. Anat. Rec. 65:15–22, 1936.
212. Franke, K. W., and E. P. Painter. A study of the toxicity and selenium content of seleniferous diets: With statistical consideration. Cereal Chem. 15:1–24, 1938.
213. Franke, K. W., and E. P. Painter. Effect of sulfur additions on seleniferous soils. Ind. Eng. Chem. 29:591–595, 1937.
214. Franke, K. W., and E. P. Painter. Selenium in proteins from toxic foodstuffs. I. Remarks on the occurrence and nature of the selenium present in a number of foodstuffs or their derived products. Cereal Chem. 13:67–70, 1936.
215. Franke, K. W., and E. P. Painter. Selenium in proteins from toxic foodstuffs. IV. The effect of feeding toxic proteins, toxic protein hydrolysates, and toxic protein hydrolysates from which the selenium has been removed. J. Nutr. 10:599–611, 1935.
216. Franke, K. W., and V. R. Potter. A new toxicant occurring naturally in certain samples of plant foodstuffs. IX. Toxic effects of orally ingested selenium. J. Nutr. 10:213–221, 1935.
217. Franke, K. W., and V. R. Potter. The ability of rats to discriminate between diets of varying degrees of toxicity. Science 83:330–332, 1936.
218. Franke, K. W., T. D. Rice, A. G. Johnson, and H. W. Schoening. Report on a Preliminary Field Survey of the So-Called "Alkali Disease" of Livestock. U.S. Department of Agriculture Circular No. 320. Washington, D.C.: U.S. Department of Agriculture, 1934. 9 pp.
219. Franke, K. W., and W. C. Tully. A new toxicant occurring naturally in certain samples of plant foodstuffs. V. Low hatchability due to deformities in chicks. Poultry Sci. 14:273–279, 1935.
220. Froom, J. D. Identification of Dimethyl Selenide as a Respiratory Product from Organisms Administered Selenium Compounds. M. S. thesis, Brookings: South Dakota State University, 1963. 45 pp.

221. Frost, D. V. Recent advances in trace elements: Emphasis on interrelationships, pp. 31-40. In Proceedings of Cornell Nutrition Conference on Feed Manufactures, 1967.
222. Frost, D. V. Selenium has great nutritional significance for man; should be cleared for feed. Feedstuffs 44(8):58-59, 1972.
223. Frost, D. V. Significance of the symposium, pp. 7-26. In O. H. Muth, Ed. Symposium: Selenium in Biomedicine. First International Symposium, Oregon State University, 1966. Westport, Conn.: AVI Publishing Co., 1967.
224. Frost, D. V. The case for selenite as a feed additive. Feedstuffs 43:12, 1971.
225. Frost, D. V. The two faces of selenium—Can selenophobia be cured? CRC Crit. Rev. Toxicol. 1:467-514, 1972.
226. Fukuyama, T., and E. J. Ordal. Induced biosynthesis of formic hydrogenlyase in iron-deficient cells of *Escherichia coli*. J. Bacteriol. 90:673-680, 1965.
227. Fuller, J. M., E. D. Beckman, M. Goldman, and L. K. Bustad. Selenium determination in human and swine tissues by x-ray emission spectometry, pp. 119-124. In O. H. Muth, Ed. Symposium: Selenium in Biomedicine. First International Symposium, Oregon State University, 1966. Westport, Conn.: AVI Publishing Co., 1967.
228. Gabbedy, B. J. Effect of selenium on wool production body weight and mortality of young sheep in western Australia. Austral. Vet. J. 47:318-322, 1971.
229. Gabbedy, B. J. Toxicity in sheep associated with the prophylactic use of selenium. Austral. Vet. J. 46:223-226, 1970.
230. Gabbedy, B. J., and J. Dickson. Acute selenium poisoning in lambs. Austral. Vet. J. 45:470-472, 1969.
231. Gabbedy, B. J., and R. B. Richards. White muscle disease in a foal. Austral. Vet. J. 46:111-112, 1970.
232. Ganje, T. J., and E. I. Whitehead. Selenium uptake by plants as affected by the forms of selenium in the soil. Proc. S. Dakota Acad. Sci. 37:85-88, 1958.
233. Ganther, H. E. Enzymic synthesis of dimethyl selenide from sodium selenite in mouse liver extracts. Biochemistry 5:1089-1098, 1966.
234. Ganther, H. E. Reduction of the selenotrisulfide derivative of glutathione to a persulfide analog by glutathione reductase. Biochemistry 10:4089-4098, 1971.
235. Ganther, H. E. Selenium: The biological effects of a highly active trace substance, pp. 211-221. In D. D. Hemphill, Ed. Trace Substances in Environmental Health. Proceedings of University of Missouri's 4th Annual Conference on Trace Substances in Environmental Health, June 23, 24, 25, 1970. Columbia: University of Missouri, 1971.
236. Ganther, H. E. Selenotrisulfides. Formation by the reaction of thiols with selenious acid. Biochemistry 7:2898-2905, 1968.
237. Ganther, H. E., and C. A. Baumann. Selenium metabolism. I. Effects of diet, arsenic and cadmium. J. Nutr. 77:210-216, 1962.
238. Ganther, H. E., and C. A. Baumann. Selenium metabolism. II. Modifying effects of sulfate. J. Nutr. 77:408-414, 1962.
239. Ganther, H. E., and C. Corcoran. Selenotrisulfides. II. Cross-linking of reduced pancreatic ribonuclease with selenium. Biochemistry 8:2557-2563, 1969.
240. Ganther, H. E., C. Goudie, M. L. Sunde, M. J. Kopecky, P. Wagner, S.-H. Oh, and W. G. Hoekstra. Selenium: Relation to decreased toxicity of methylmercury added to diets containing tuna. Science 175:1122-1124, 1972.
241. Ganther, H. E., O. A. Levander, and C. A. Baumann. Dietary control of selenium volatilization in the rat. J. Nutr. 88:55-60, 1966.
242. Ganther, H. E., P. A. Wagner, M. L. Sunde, and W. G. Hoekstra. Protective effects of selenium against heavy metal toxicities, pp. 247-259. In D. D. Hemphill, Ed. Trace Substances in Environmental Health. Proceedings of the University of Missouri's

References

6th Annual Conference on Trace Substances in Environmental Health, June 13, 14, and 15, 1972, Columbia, Missouri. Columbia: University of Missouri, 1973.

243. Gardiner, M. R., J. Armstrong, H. Fels, and R. H. Glencross. A preliminary report on selenium and animal health in Western Australia. Austral. J. Exp. Agric. Animal Husb. 2:261–269, 1962.

244. Gardner, R. W., and D. E. Hogue. Milk levels of selenium and vitamin E related to nutritional muscular dystrophy in the suckling lamb. J. Nutr. 93:418–424, 1967.

245. Garrow, J. S., and C. P. Douglas. A rapid method for assessing intrauterine growth by radioactive selenomethionine uptake. J. Obstet. Gynaec. Brit. Commonw. 75:1034–1039, 1968.

246. Geering, H. R., E. E. Cary, L. H. P. Jones, and W. H. Allaway. Solubility and redox criteria for the possible forms of selenium in soils. Soil Sci. Soc. Amer. Proc. 32:35–40, 1968.

247. Gile, P. L., and H. W. Lakin. Effect of different soil colloids on the toxicity of sodium selenite to millet. J. Agric. Res. 63:560–581, 1941.

248. Gile, P. L., and H. W. Lakin. The influence of soil colloids on the toxicities of sodium selenate and sodium selenite for millet. Soil Sci. Soc. Amer. Proc. 3:92–93, 1938.

249. Gile, P. L., H. W. Lakin, and H. G. Byers. Effect of different soil colloids and whole soils on the toxicity of sodium selenate to millet. J. Agric. Res. 57:1–20, 1938.

250. Giordano, W. C. One application treatment for tinea versicolor. J. Med. Soc. N.J. 60:186–187, 1963.

251. Gissel-Nielsen, G. Influence of pH and texture of the soil on plant uptake of added selenium. J. Agric. Food Chem. 19:1165–1167, 1971.

252. Gissel-Nielsen, G. Selenium content of some fertilizers and their influence on uptake of selenium in plants. J. Agric. Food Chem. 19:564–566, 1971.

253. Gissel-Nielsen, G. Uptake and distribution of added selenite and selenate by barley and red clover as influenced by sulphur. J. Sci. Food Agric. 24:649–655, 1973.

254. Gissel-Nielsen, G., and B. Bisbjerg. The uptake of applied selenium by agricultural plants. 2. The utilization of various selenium compounds. Plant Soil 32:382–396, 1970.

255. Gleit, C. E., and W. D. Holland. Use of electrically excited oxygen for the low temperature decomposition of organic substances. Anal. Chem. 34:1454–1457, 1962.

256. Glenn, M. W., J. L. Martin, and L. M. Cummins. Sodium selenate toxicosis: The distribution of selenium within the body after prolonged feeding of toxic quantities of sodium selenate to sheep. Amer. J. Vet. Res. 25:1495–1499, 1964.

257. Glover, J. R. Selenium and its industrial toxicology. Ind. Med. Surg. 39:50–54, 1970.

258. Glover, J. R. Some medical problems concerning selenium in industry. Trans. Assoc. Ind. Med. Officers. 4:94–96, 1954.

259. Godwin, K. O., and C. N. Fuss. The entry of selenium into rabbit protein following the administration of $Na_2{}^{75}SeO_3$. Austral. J. Biol. Sci. 25:865–871, 1972.

260. Goel, Y., J. Sims, and J. A. Pittmen. Mediastinum scanning with ^{75}Se-selenomethionine. J. Nucl. Med. 12:644–645, 1971.

261. Goldman, M., and E. D. Beckman. X-Ray Emission Quantitation of Trace Elements in Biomedical Research. U.S. Atomic Energy Commission Report No. UCD-472-210. Davis: University of California, Radiobiology Laboratory, 1967. 26 pp.

262. Goldschmidt, H., and A. M. Kligman. Increased sebum secretion following selenium sulfide shampoos. Acta Derm.-Venereol. 48:489–491, 1968.

263. Goldschmidt, V. M., and L. W. Strock. Zur Geochemie des Selen II. Nachr. Ges. Wiss. Göttingen Math.-Physik. Klasse 1:123–142, 1935.

264. Gortner, R. A., Jr., and H. B. Lewis. Retention and excretion of selenium after the

administration of sodium selenite to white rats. J. Pharmacol. Exp. Therap. 67: 358-364, 1939.
265. Goto, T., and M. Fujino. Selenium content in foods and its analysis. Eiyo To Shokuryo 20:311-313, 1967.
266. Grant, A. B. Pasture top-dressing with selenium. N.Z. J. Agric. Res. 8:681-690, 1965.
267. Grant, A. B., and G. F. Wilson. Selenium content of milk from cows given sodium selenate. N.Z. J. Agric. Res. 11:733-736, 1968.
268. Grant, C. A., B. Thafvelin, and R. Christell. Retention of selenium by pig tissues. Acta Pharmacol. Toxicol. 18:285-297, 1961.
269. Gries, C. L., and M. L. Scott. Pathology of selenium deficiency in the chick. J. Nutr. 102:1287-1296, 1972.
270. Groce, A. W., E. R. Miller, K. K. Keahey, D. E. Ullrey, and D. J. Ellis. Selenium supplementation of practical diets for growing-finishing swine. J. Animal Sci. 32: 905-911, 1971.
271. Groce, A. W., E. C. Rossman, and D. E. Ullrey. Selenium levels of Michigan-grown corn. J. Animal Sci. 35:1104-1105, 1972.
272. Groce, A. W., D. E. Ullrey, E. R. Miller, D. J. Ellis, and K. K. Keahey. Selenium and vitamin E in practical swine diets. J. Animal Sci. 33:230-231, 1971.
273. Grover, R. W. Diffuse hair loss associated with selenium (Selsun) sulfide shampoo. J.A.M.A. 160:1397-1398, 1956.
274. Gruenwald, P. Malformations caused by necrosis in the embryo. Illustrated by the effect of selenium compounds on chick embryos. Amer. J. Path. 34:77-103, 1958.
275. Gunn, S. A., and T. C. Gould. Cadmium and other mineral elements, pp. 377-481. In A. D. Johnson, W. R. Gomes, and N. D. Vandemark, Eds. The Testis. Vol. III. Influencing Factors. New York: Academic Press, Inc., 1970.
276. Gunn, S. A., and T. C. Gould. Specificity of response in relation to cadmium, zinc and selenium, pp. 395-413. In O. H. Muth, Ed. Symposium: Selenium in Biomedicine. First International Symposium, Oregon State University, 1966. Westport, Conn.: AVI Publishing Co., 1967.
277. Gunn, S. A., and T. C. Gould. Vasculature of the testes and adnexa, pp. 117-142. In R. O. Greep and E. B. Astwood, Eds. Handbook of Physiology. Section 7: Endocrinology. Vol. 5. Male Reproductive System. Washington, D.C.: American Physiological Society, 1975.
278. Gunn, S. A., T. C. Gould, and W. A. D. Anderson. Incorporation of selenium into spermatogenic pathway in mice. Proc. Soc. Exp. Biol. Med. 124:1260-1263, 1967.
279. Gunn, S. A., T. C. Gould, and W. A. D. Anderson. Mechanisms of zinc, cysteine and selenium protection against cadmium-induced vascular injury to mouse testis. J. Reprod. Fertil. 15:65-70, 1968.
280. Gunn, S. A., T. C. Gould, and W. A. D. Anderson. Protective effect of thiol compounds against cadmium-induced vascular damage to testis. Proc. Soc. Exp. Biol. Med. 122:1036-1039, 1966.
281. Gunn, S. A., T. C. Gould, and W. A. D. Anderson. Protective effects of estrogen against vascular damage to the testis caused by cadmium. Proc. Soc. Exp. Biol. Med. 119:901-905, 1965.
282. Gunn, S. A., T. C. Gould, and W. A. D. Anderson. Selectivity of organ response to cadmium injury and various protective measures. J. Path. Bacteriol. 96:89-96, 1968.
283. Gunn, S. A., T. C. Gould, and W. A. D. Anderson. Specificity in protection against lethality and testicular toxicity from cadmium. Proc. Soc. Exp. Biol. Med. 128: 591-595, 1968.

284. Gunn, S. A., T. C. Gould, and W. A. D. Anderson. The selective injurious response of testicular and epididymal blood vessels to cadmium and its prevention by zinc. Amer. J. Path. 42:685–702, 1963.
285. Gunn, S. A., T. C. Gould, and W. A. D. Anderson. Zinc protection against cadmium injury to rat testis. Arch. Path. 71:274–281, 1961.
286. Gutenmann, W. H., and D. J. Lisk. Determination of selenium in oats by oxygen flask combustion. J. Agric. Food Chem. 9:488–489, 1961.
287. Hadjimarkos, D. M. Effect of trace elements on dental caries. Adv. Oral Biol. 3: 253–292, 1968.
288. Hadjimarkos, D. M. Selenium content of human milk: Possible effects on dental caries. J. Pediat. 63:273–275, 1963.
289. Hadjimarkos, D. M. Selenium in man. N.Z. Med. J. 72:205–206, 1970.
290. Hadjimarkos, D. M. Selenium toxicity: Effect of fluoride. Experientia 25:485–486, 1969.
291. Hadjimarkos, D. M. Urinary selenium and dental caries. Nature 188:677, 1960.
292. Hadjimarkos, D. M., and C. W. Bonhorst. Selenium content of fresh eggs. 202:296, 1964.
293. Hadjimarkos, D. M., and C. W. Bonhorst. The selenium content of eggs, milk, and water in relation to dental caries in children. J. Pediat. 59:256–259, 1961.
294. Hadjimarkos, D. M., C. W. Bonhorst, and J. J. Mattice. The selenium concentration in placental tissue and fetal cord blood. J. Pediat. 54:296–298, 1959.
295. Hadjimarkos, D. M., and T. R. Shearer. Selenium concentration in human saliva. Amer. J. Clin. Nutr. 24:1210–1211, 1971.
296. Hadjimarkos, D. M., C. A. Storvick, and L. F. Remmert. Selenium and dental caries. An investigation among school children of Oregon. J. Pediat. 40:451–455, 1952.
297. Hall, R. H., S. Laskin, P. Frank, E. A. Maynard, and C. H. Hodge. Preliminary observations on toxicity of elemental selenium. Arch. Ind. Hyg. Occup. Med. 4: 458–464, 1951.
298. Hall, R. J., and P. L. Gupta. The determination of very small amounts of selenium in plant samples. Analyst 94:292–299, 1969.
299. Haller, W. A., L. A. Rancitelli, and J. A. Cooper. Instrumental determination of trace elements in plant tissue by neutron activation analysis and Ge (Li) gamma-ray spectrometry. J. Agric. Food Chem. 16:1036–1040, 1968.
300. Halverson, A. E., and K. J. Monty. An effect of dietary sulfate on selenium poisoning in the rat. J. Nutr. 70:100–102, 1960.
301. Halverson, A. W., P. L. Guss, and O. E. Olson. Effect of sulfur salts on selenium poisoning in the rat. J. Nutr. 77:459–464, 1962.
302. Halverson, A. W., C. M. Hendrick, and O. E. Olson. Observations on the protective effect of linseed oil meal and some extracts against chronic selenium poisoning in rats. J. Nutr. 56:51–60, 1955.
303. Halverson, A. W., I. S. Palmer, and P. L. Guss. Toxicity of selenium to post-weanling rats. Toxicol. Appl. Pharmacol. 9:477–484, 1966.
304. Hambidge, K. M., M. L. Franklin, and M. A. Jacobs. Hair chromium concentration. Effects of sample washing and external environment. Amer. J. Clin. Nutr. 25: 384–389, 1972.
305. Hamilton, A., and H. L. Hardy. Selenium, pp. 188–192. In Industrial Toxicology. New York: Paul B. Hoeber, Inc., 1949.
306. Hamilton, J. W., and O. A. Beath. Amount and chemical form of selenium in vegetable plants. J. Agric. Food Chem. 12:371–374, 1964.
307. Handley, R. Fluorescent x-ray determination of selenium in plant material. Anal. Chem. 32:1719–1720, 1960.

308. Handley, R., and C. M. Johnson. Microdetermination of selenium in biological materials. Anal. Chem. 31:2105-2106, 1959.
309. Handreck, K. A., and K. D. Godwin. Distribution in the sheep of selenium derived from ^{75}Se-labelled ruminal pellets. J. Nutr. 79:493-502, 1970.
310. Handreck, K. A., and K. D. Godwin. Distribution in the sheep of selenium derived from ^{75}Se-labelled ruminal pellets. Austral. J. Agric. Res. 21:71-84, 1970.
311. Hansson, E., and M. Blau. Incorporation of ^{75}Se Se-selenomethionine into pancreatic juice proteins *in vivo*. Biochem. Biophys. Res. Commun. 13:71-74, 1963.
312. Hansson, E., and S.-O. Jacobsson. Uptake of (^{75}Se) selenomethionine in the tissues of the mouse studied by whole-body autoradiography. Biochim. Biophys. Acta 115:285-293, 1966.
313. Harr, J. R., J. F. Bone, I. J. Tinsley, P. H. Weswig, and R. S. Yamamoto. Selenium toxicity in rats. II. Histopathology, pp. 153-178. In O. H. Muth, Ed. Symposium: Selenium in Biomedicine. First International Symposium, Oregon State University, 1966. Westport, Conn.: AVI Publishing Co., 1967.
314. Harr, J. R., J. H. Exon, P. H. Weswig, and P. D. Whanger. Relationship of dietary selenium concentration; chemical cancer induction; and tissue concentration of selenium in rats. Clin. Toxicol. 6:487-495, 1973.
315. Harr, J. R., J. H. Exon, P. D. Whanger, and P. H. Weswig. Effect of dietary selenium on N-2-fluorenyl-acetamide (FAA)-induced cancer in vitamin E supplemented, selenium depleted rats. Clin. Toxicol. 5:187-194, 1972.
316. Harr, J. R., and O. H. Muth. Selenium poisoning in domestic animals and its relationship to man. Clin. Toxicol. 5:175-186, 1972.
317. Harrison, P. R., K. A. Rahn, R. Dams, J. A. Robbins, J. W. Winchester, S. S. Brar, and D. M. Nelson. Areawide trace metal concentrations measured by multielement neutron activation analysis. A one day study in northwest Indiana. J. Air Pollut. Control Assoc. 21:563-570, 1971.
318. Harrison, W. W., G. C. Clemena, and C. W. Magee. Forensic applications of spark source mass spectrometry. J. Assoc. Off. Anal. Chemists 54:929-936, 1971.
319. Hartley, W. J. Levels of selenium in animal tissues and methods of selenium administration, pp. 79-96. In O. H. Muth, Ed. Symposium: Selenium in Biomedicine. First International Symposium, Oregon State University, 1966. Westport, Conn.: AVI Publishing Co., 1967.
320. Hartley, W. J., and A. B. Grant. A review of selenium responsive diseases of New Zealand livestock. Fed. Proc. 20:679-688, 1961.
321. Hartley, W. J., A. B. Grant, and C. Drake. Recent advances in selenium and animal health. Massey Agric. Coll. Sheepfarming Ann. 1960: 43-48.
322. Hashimoto, Y., J. Y. Hwang, and S. Yanagisawa. Possible source of atmospheric pollution of selenium. Environ. Sci. Technol. 4:157-158, 1970.
323. Hashimoto, Y., and J. W. Winchester. Selenium in the atmosphere. Environ. Sci. Technol. 1:338-340, 1967.
324. Hawley, J. E., and I. Nichol. Selenium in some Canadian sulfides. Econ. Geol. 54: 608-628, 1959.
325. Haynie, T. P., W. K. Otte, and J. C. Wright. Visualization of hyperfunctioning parathyroid adenoma using Se^{75} selenomethionine and the photoscanner. J. Nucl. Med. 5:710-714, 1964.
326. Heinrich, M., Jr., and F. E. Kelsey. Selenium metabolism: Loss of selenium from mouse tissues on heating. Fed. Proc. 13:364, 1954.
327. Heinrich, M., Jr., and F. E. Kelsey. Studies on selenium metabolism. The distribution of selenium in the tissues of the mouse. J. Pharmacol. Exp. Ther. 114:28-32, 1955.

328. Henschler, D., and U. Kirschner. Zur Resorption und Toxizität von Selensulfid. Arch. Toxikol. 24:341–344, 1969.
329. Herigstad, R. R., C. K. Whitehair, and O. E. Olson. Inorganic and organic selenium toxicosis in young swine: Comparison of pathologic changes with those in swine with vitamin E-selenium deficiency. Amer. J. Vet. Res. 34:1227–1238, 1973.
330. Herrera, N. E., R. D. Gonzalez, and R. N. Kranwinkel. Further investigation of role of selenomethionine Se^{75} in diagnosis of lymphoma. Lahey Clin. Bull. 17:43–49, 1968.
331. Herrera, N. E., R. Gonzalez, R. D. Schwartz, A. M. Diggs, and J. Belsky. ^{75}Se-methionine as diagnostic agent in malignant lymphoma. J. Nucl. Med. 6:792–804, 1965.
332. Hersle, K. Selenium sulphide treatment of tinea versicolor. Acta Dermato-Venereol. 51:476–478, 1971.
333. Hidiroglou, M., R. B. Carson, and G. A. Brossard. Influence of selenium on the selenium contents of hair and on the incidence of nutritional muscular disease in beef cattle. Can. J. Anim. Sci. 45:197–202, 1965.
334. Hidiroglou, M., D. P. Heaney, and K. J. Jenkins. Metabolism of inorganic selenium in rumen bacteria. Can. J. Physiol. Pharmacol. 46:229–232, 1968.
335. Hidiroglou, M., I. Hoffman, and K. J. Jenkins. Selenium distribution and radiotocopherol metabolism in the pregnant ewe and fetal lamb. Can. J. Physiol. Pharmacol. 47:953–962, 1969.
336. Hidiroglou, M., K. J. Jenkins, and I. Hoffman. Teneurs en sélénium dans les tissus des ruminants. Ann. Biol. Anim. Biochim. Biophys. 11:695–704, 1971.
337. Higgs, D. J., V. C. Morris, and O. A. Levander. Effect of cooking on selenium content of foods. J. Agric. Food Chem. 20:678–680, 1972.
338. Hill, J. C. H. Reversal of selenium toxicity in chicks by mercury and cadmium. J. Nutr. 104:593–598, 1974.
339. Hill, M. K., S. D. Walker, and A. G. Taylor. Effects of "marginal" deficiencies of copper and selenium on growth and productivity of sheep. N.Z. J. Agric. Res. 12:261–270, 1969.
340. Hoffman, J. L., K. P. McConnell, and D. R. Carpenter. Aminoacylation of *Escherichia coli* methionine tRNA by selenomethionine. Biochim. Biophys. Acta 199:531–534, 1970.
341. Hogue, D. E., J. F. Procter, R. G. Warner, and J. K. Loosli. Relation of selenium, vitamin E, and an unidentified factor to muscular dystrophy (stiff-lamb or white-muscle disease) in the lamb. J. Anim. Sci. 21:25–29, 1962.
342. Holker, J. R., and J. B. Speakman. The action of selenium dioxide on wool. J. Appl. Chem. 8:1–3, 1958.
343. Holmberg, R. E., Jr., and V. H. Ferm. Interrelationships of selenium, cadmium and arsenic in mammalian teratogenesis. Arch. Environ. Health 18:873–877, 1969.
344. Hopkins, L. L., Jr., and A. S. Majaj. Selenium in human nutrition, pp. 203–214. In O. H. Muth, Ed. Symposium: Selenium in Biomedicine. First International Symposium, Oregon State University, 1966. Westport, Conn.: AVI Publishing Co., 1967.
345. Hopkins, L. L., Jr., A. L. Pope, and C. A. Baumann. Distribution of microgram quantities of selenium in the tissues of the rat and effects of previous selenium intake. J. Nutr. 88:61–65, 1966.
346. Horn, M. J., and D. B. Jones. Isolation from *Astragalus pectinatus* of a crystalline amino acid complex containing selenium and sulfur. J. Biol. Chem. 139:649–660, 1941.
347. Horn, M. J., and D. B. Jones. Isolation of a crystalline selenium-containing organic compound from plant material. J. Amer. Chem. Soc. 62:234, 1940. (letter)

348. Hoste, J. Diaminobenzidine as a reagent for vanadium and selenium. Anal. Chim. Acta 2:402–408, 1948.
349. Howard, J. H. Control of geochemical behavior of selenium in natural waters by adsorption on hydrous ferric oxides, pp. 485–495. In D. D. Hemphill, Ed. Trace Substances in Environmental Health. Proceedings of University of Missouri's 5th Annual Conference on Trace Substances in Environmental Health, June 29–July 1, 1971. Columbia: University of Missouri, 1972.
350. Howard, J. H. Source of selenium in sediments of western Great Plains: Late cretaceous volcanic centers of west-central Montana. Geol. Soc. Amer. Abstr. 5:485, 1973.
351. Huber, R. E., and R. S. Criddle. Comparison of the chemical properties of selenocysteine and selenocystine with their sulfur analogs. Arch. Biochem. Biophys. 122:164–173, 1967.
352. Huber, R. E., and R. S. Criddle. The isolation and properties of β-galactosidase from *Escherichia coli* grown on sodium selenate. Biochim. Biophys. Acta 141:587–599, 1967.
353. Huber, R. E., I. H. Segel, and R. S. Criddle. Growth of *Escherichia coli* on selenate. Biochim. Biophys. Acta 141:573–586, 1967.
354. Hunter, W. C., and J. M. Roberts. Glomerular changes in the kidneys of rabbits and monkeys induced by uranium nitrate, mercuric chloride and potassium bichromate. Amer. J. Path. 8:665–688, 1932.
355. Hurd-Karrer, A. M. Comparative toxicity of selenates and selenites to wheat. Amer. J. Botany 24:720–728, 1937.
356. Hurd-Karrer, A. M. Factors affecting the absorption of selenium from soils by plants. J. Agric. Res. 50:413–427, 1935.
357. Hurd-Karrer, A. M. Inhibition of selenium injury to wheat plants by sulfur. Science 78:560, 1933.
358. Hurd-Karrer, A. M. Relation of sulphate to selenium absorption by plants. Amer. J. Botany 25:666–675, 1938.
359. Hurd-Karrer, A. M. Selenium absorption by crop plants as related to their sulphur requirement. J. Agric. Res. 54:601–608, 1937.
360. Hurd-Karrer, A. M. Selenium injury to wheat plants and its inhibition by sulphur. J. Agric. Res. 49:343–357, 1934.
361. Hurd-Karrer, A. M., and M. H. Kennedy. Inhibiting effect of sulphur in selenized soil on toxicity of wheat to rats. J. Agric. Res. 52:933–942, 1936.
362. Hurt, H. D., E. E. Cary, W. H. Allaway, and W. J. Visek. Effect of dietary selenium on the survival of rats exposed to chronic whole body irradiation. J. Nutr. 101:363–366, 1971.
363. Hurt, H. D., E. E. Cary, and W. J. Visek. Growth, reproduction, and tissue concentrations of selenium in the selenium-depleted rats. J. Nutr. 101:761–766, 1971.
364. Ishibashi, M. Minute elements in sea water. Rec. Oceanogr. Works Jap. 1(1):88–92, 1953.
365. Ishibashi, M., T. Shigematsu, and Y. Nakagawa. Determination of selenium in sea water. Rec. Oceanogr. Works Jap. 1(2):44–48, 1953.
366. Jacobs, A. L. The Isolation and Identification of a Seleno Amino Acid from Corn. Ph.D. thesis, New York: Columbia University, 1962. 50 pp.
367. Jacobsson, S. O. Uptake of Se^{75} in tissues of sheep after administration of a single dose of Se^{75}-sodium selenite, Se^{75}-selenomethionine, or Se^{75}-selenocystine. Acta Vet. Scand. 7:303–320, 1966.
368. Jacobsson, S. O., and H. E. Oksanen. The placental transmission of selenium in sheep. Acta Vet. Scand. 7:66–76, 1966.

References

369. Jaffe, W. G., M. D. Ruphael, M. C. Mondragon, and M. A. Cuevas. Estudio clínico y bioquímico en niños escolares de una zona selenífera. Arch. Latinoamer. Nutr. 22:595-611, 1972.
370. Jenkins, K. J. Evidence for the absence of selenocystine and selenomethionine in the serum proteins of chicks administered selenite. Can. J. Biochem. 46:1417-1425, 1968.
371. Jenkins, K. J., and M. Hidiroglou. Comparative uptake of selenium by low cystine and high cystine proteins. Can. J. Biol. 49:468-472, 1971.
372. Jenkins, K. J., and M. Hidiroglou. The incorporation of ^{75}Se-selenite into dystrophogenic pasture grass. The chemical nature of the seleno compounds formed and their availability to young ovine. J. Biochem. 45:1027-1039, 1967.
373. Jenkins, K. J., M. Hidiroglou, and J. F. Ryan. Intravascular transport of selenium by chick serum proteins. Can. J. Physiol. Pharmacol. 47:459-467, 1969.
374. Jensen, L. S. Selenium deficiency and impaired reproduction in Japanese quail. Proc. Soc. Exp. Biol. Med. 128:970-972, 1968.
375. Jensen, L. S. Vitamin E and essential fatty acids in avian reproduction. Fed. Proc. 27:914-919, 1968.
376. Jensen, L. S., and J. McGinnis. Influence of selenium, antioxidants and type of yeast on Vitamin E deficiency in the adult chicken. J. Nutr. 72:23-28, 1960.
377. Jensen, L. S., E. D. Walter, and J. S. Dunlap. Influence of dietary vitamin E and selenium on distribution of Se in the chick. Proc. Soc. Exp. Biol. Med. 112:899-901, 1963.
378. Johnson, H. Determination of selenium in solid waste. Environ. Sci. Technol. 4:850-853, 1970.
379. Johnson, R. R., and E. I. Whitehead. Growth and selenium content of wheat plants as related to the selenite selenium content of soil. Proc. S. Dakota Acad. Sci. 30:130-136, 1951.
380. Johnston, W. K. The Effect of Selenium on Chemical Carcinogenicity in the Rat. M.S. thesis, Corvallis: Oregon State University, 1974. 57 pp.
381. Jones, C. O. The physiological effects of selenium compounds with relation to their action of glycogen and sugar derivatives in the tissues. Biochem. J. 4:405-419, 1909.
382. Jones, G. B., and K. O. Godwin. Distribution of radioactive selenium in mice. Nature 196:1294-1296, 1962.
383. Jones, G. B., and K. O. Godwin. Studies on the nutritional role of selenium. I. The distribution of radioactive selenium in mice. Austral. J. Agric. Res. 14:716-723, 1963.
384. Kahn, H. L., and J. E. Schallis. Improvement of detection limits for arsenic, selenium, and other elements with an argon–hydrogen flame. Atomic Absorption Newslett. 7:5-9, 1968.
385. Kaplan, E., and M. Domingo. 75 Se-selenomethionine in hepatic focal lesions. Semin. Nucl. Med. 2:139-149, 1972.
386. Kar, A. B., and R. P. Das. The nature of protective action of selenium on cadmium-induced degeneration of the rat testis. Proc. Nat. Inst. Sci. India 29B:297-305, 1963.
387. Kar, A. B., R. P. Das, and F. N. I. Mukerji. Prevention of cadmium-induced changes in the gonads of rat by zinc and selenium—A study in antagonism between metals in the biological system. Proc. Nat. Inst. Sci. India 26B:40-50, 1960.
388. Kawashima, T., and M. Tanaka. Determination of submicrogram amounts of selenium (IV) by means of the catalytic reduction of 1,4,6,11-tetrazanaphthacene. Anal. Chim. Acta 40:137-143, 1968.
389. Kerdel-Vegas, F., F. Wagner, P. B. Russell, N. H. Grant, H. E. Alburn, D. E. Clark, and J. A. Miller. Structure of the pharmacologically active factor in the seeds of *Lecythis ollaria*. Nature 205:1186-1187, 1965.

390. Kiermeier, F., and W. Wigand. Selengehalt von Milch und Milchpulver. Z. Lebensm.-Unters. -Forsch. 139:205-211, 1969.
391. Kifer, R. R., and W. L. Payne. Selenium content of fish meals. Feedstuffs 40(35):32, 1968.
392. Kifer, R. R., W. L. Payne, and M. E. Ambrose. Selenium content of fish meals. II. Feedstuffs 41(51):24-25, 1969.
393. Kimmerle, G. Vergleichende Untersuchungen der Inhalations Toxcität von Schwefel-, Selen-, and Tellurhexafluorid. Arch. Toxicol. 18:140-144, 1960.
394. Kirkbright, G. F., and J. H. Yoe. A new spectrophotometric method for the determination of microgram amounts of selenium. Anal. Chem. 35:808-811, 1963.
395. Klayman, D. L. Selenium compounds as potential chemotherapeutic agents, pp. 727-761. In D. L. Klayman and W. H. H. Günther, Eds. Organic Selenium Compounds: Their Chemistry and Biology. New York: Wiley-Interscience, 1973.
396. Klayman, D. L., and W. H. H. Günther, Eds. Organic Selenium Compounds: Their Chemistry and Biology. New York: Wiley-Interscience, 1973. 1188 pp.
397. Klein, A. K. Report on selenium. J. Assoc. Off. Agric. Chemists 24:363-380, 1941.
398. Klevay, L. M. Hair as a biopsy material. I. Assessment of zinc nutriture. Amer. J. Clin. Nutr. 23:284-289, 1970.
399. Klevay, L. M. Hair as biopsy material. II. Assessment of copper nutriture. Amer. J. Clin. Nutr. 23:1194-1202, 1970.
400. Klug, H. L., R. D. Harshfield, R. M. Pengra, and A. L. Moxon. Methionine and selenium toxicity. J. Nutr. 48:409-420, 1952.
401. Klug, H. L., D. F. Petersen, and A. L. Moxon. The toxicity of selenium analogues of cystine and methionine. Proc. S. Dakota Acad. Sci. 28:117-120, 1949.
402. Knight, R. A., A. A. Christie, C. R. Orton, and J. Robertson. Metabolic balance of zinc, copper, cadmium, iron, molybdenum and selenium in young New Zealand women. Brit. J. Nutr. 30:181-188, 1973.
403. Knight, S. H., and O. A. Beath. The Occurrence of Selenium and Seleniferous Vegetation in Wyoming. Part I. The Rocks and Soils of Wyoming and Their Relations to the Selenium Problem. Part II. Seleniferous Vegetation of Wyoming. Wyoming Agricultural Experiment Station Bulletin No. 221. Laramie: University of Wyoming Agricultural Experiment Station, 1937. 64 pp.
404. Kovalskii, V. V., and V. V. Ermakov. The Biological Importance of Selenium. (Translated by S. J. Wilson) Boston Spa, Yorkshire, England: National Lending Library of Science and Technology, 1970. 143 pp.
405. Kovalskii, V. V., S. V. Letunova, and R. D. Altynbayeva. Genetic transformation of selenium resistance and selenium reductase formation in *Bacillus megaterium*. Dokl. Akad. Nauk SSSR 194:1429-1432, 1970. (in Russian)
406. Ku, P. K., W. T. Ely, A. W. Groce, and D. E. Ullrey. Natural dietary selenium, α-tocopherol and effect on tissue selenium. J. Anim. Sci. 34:208-211, 1972.
407. Ku, P. K., E. R. Miller, R. C. Wahlstrom, A. W. Groce, J. P. Hitchcock, and D. E. Ullrey. Selenium supplementation of naturally high selenium diets for swine. J. Anim. Sci. 37:501-505, 1973.
408. Ku, P. K., E. R. Miller, R. C. Wahlstrom, A. W. Groce, and D. E. Ullrey. Supplementation of naturally high Se diets. J. Anim. Sci. 35:218-219, 1972.
409. Kubota, J., W. H. Allaway, D. L. Carter, E. E. Cary, and V. A. Lazar. Selenium in crops in the United States in relation to selenium-responsive diseases of animals. J. Agric. Food Chem. 15:448-453, 1967.
410. Kuchel, R. E., and R. A. Buckley. The provision of selenium to sheep by means of heavy pellets. Austral. J. Agric. Res. 20:1099-1107, 1969.

References

411. Kury, G., L. H. Rev-Kury, and R. J. Crosby. The effect of selenous acid on the hematopoietic system of chicken embryos. Toxicol Appl. Pharmacol. 11:449–458, 1967.
412. Kuttler, K. L., and D. W. Marble. Prevention of white muscle disease in lambs by oral and subcutaneous administration of selenium. Amer. J. Vet. Res. 21:437–440, 1960.
413. Kuttler, K. L., D. W. Marble, and C. Blincoe. Serum and tissue residues following selenium injections in sheep. Amer. J. Vet. Res. 22:422–428, 1961.
414. Lakin, H. W., and H. G. Byers. Selenium in wheat and wheat products. Cereal Chem. 18:73–78, 1941.
415. Lakin, H. W., and H. G. Byers. Selenium Occurrence in Certain Soils in the United States with a Discussion of Related Topics. Sixth Report. U.S. Department of Agriculture Technical Bulletin No. 783. Washington, D.C.: U.S. Department of Agriculture, 1941. 26 pp.
416. Lakin, H. W., and D. F. Davidson. The relation of the geochemistry of selenium to its occurrence in soils, pp. 27–56. In O. H. Muth, Ed. Symposium: Selenium in Biomedicine. First International Symposium, Oregon State University, 1966. Westport, Conn.: AVI Publishing Co., 1967.
417. Lakin, H. W., and A. R. Trites, Jr. The behavior of selenium in the zone of oxidation, pp. 113–124. In Symposium de Exploración Geoquímica. Vol. I. International Geological Congress, 20th Session, Mexico City, 1956. Mexico City: Twentieth International Geological Congress, 1958.
418. Lamand, M. Lésions biochimiques dans la myopathie du veau par carence en sélénium. C. R. Acad. Sci. 270D:417–420, 1970.
419. Lambert, J. P. F., O. Levander, L. Argrett, and R. E. Simpson. Neutron activation analysis of selenium in biological samples. J. Assoc. Off. Anal. Chemists 52:915–917, 1960.
420. Lambourne, D. A., and R. W. Mason. Mortality in lambs following overdosing with sodium selenite. Austral. Vet. J. 45:208, 1969.
421. Landauer, W. Studies on the creeper fowl. XIII. The effect of selenium and the asymmetry of selenium-induced malformations. J. Exp. Zool. 83:431–443, 1940.
422. Lansche, A. M. Selenium and Tellurium—A Materials Survey. U.S. Bureau of Mines Information Circular No. 8340. Washington, D.C.: U.S. Department of the Interior, 1967. 56 pp.
423. Lee, P., and J. S. Garrow. A clinical evaluation of the selenomethionine uptake test. J. Obstet. Gynaec. Brit. Commonw. 77:983–986, 1970.
424. Lemley, R. E. Selenium poisoning in the human. A preliminary case report. J.-Lancet 60:528–531, 1940.
425. Lemley, R. E., and M. P. Merryman. Selenium poisoning in the human. J.-Lancet 61:435–438, 1941.
426. Leonard, R. D., and R. H. Burns. A preliminary study of selenized wool. J. Animal Sci. 14:446–457, 1955.
427. Lester, R. L., and J. A. DeMoss. Effect of molybdate and selenite on formate and nitrate metabolism in *Escherichia coli*. J. Bacteriol. 105:1006–1014, 1971.
428. Letunova, S. V. Geochemical ecology of soil micro-organisms, pp. 432–437. In C. F. Mills, Ed. Trace Element Metabolism in Animals. Proceedings of WAAP/IBP International Symposium, Aberdeen, Scotland, July 1969. Edinburgh: E. & S. Livingstone, Ltd., 1970.
429. Levan, N. E. Selenium sulfide suspension in the treatment of tinea versicolor. Arch. Derm. 75:128–129, 1957.
430. Levander, O. A., and C. A. Baumann. Selenium metabolism. V. Studies on the distribution of selenium in rats given arsenic. Toxicol. Appl. Pharmacol. 9:98–105, 1966.

431. Levander, O. A., and C. A. Baumann. Selenium metabolism. VI. Effect of arsenic on the excretion of selenium in the bile. Toxicol. Appl. Pharmacol. 9:106-115, 1966.
432. Levander, O. A., and V. C. Morris. Interactions of methionine, vitamin E, and antioxidants in selenium toxicity in the rat. J. Nutr. 100:1111-1118, 1970.
433. Levander, O. A., V. C. Morris, and D. J. Higgs. Acceleration of thiol-induced swelling of rat liver mitochondria by selenium. Biochemistry 12:4586-4590, 1973.
434. Levander, O. A., V. C. Morris, and D. J. Higgs. Selenium as a catalyst for the reduction of cytochrome c by glutathione. Biochemistry 12:4591-4595, 1973.
435. Levander, O. A., V. C. Morris, and D. J. Higgs. Selenium catalysis of swelling of rat liver mitochondria and reduction of cytochrome c by sulfur compounds. Adv. Exp. Biol. Med. 48:405-423, 1974.
436. Levander, O. A., M. L. Young, and S. A. Meeks. Studies on the binding of selenium by liver homogenates from rats fed diets containing either casein or casein plus linseed oil meal. Toxicol. Appl. Pharmacol. 16:79-87, 1970.
437. Levine, R. J., and R. E. Olson. Blood selenium in Thai children with protein-calorie malnutrition. Proc. Soc. Exp. Biol. Med. 134:1030-1034, 1970.
438. Levine, V. E. The effect of selenium compounds upon growth and germination in plants. Amer. J. Botany 12:82-90, 1925.
439. Levine, V. E. The reducing properties of microörganisms with special reference to selenium compounds. J. Bacteriol. 10:217-262, 1925.
440. Levshin, B. I. Changes in the activity of blood serum lactate-dehydrogenase isoenzymes of rats during experimental pharmacotherapy of toxic hepatitis with selenium and thiazolidine drugs. Farmakol. Toksikol. 35:195-198, 1972. (in Russian)
441. Lewis, B. G., C M. Johnson, and T. C. Broyer. Cleavage of Se-methylselenomethionine selenonium salt by cabbage leaf enzyme fraction. Biochim. Biophys. Acta 237:603-605, 1971.
442. Lewis, B. G., C. M. Johnson, and C. C. Delwiche. Release of volatile selenium compounds by plants. Collection procedures and preliminary observations. J. Agric. Food Chem. 14:638-640, 1966.
443. Lewis, H. B., J. Schultz, and R. H. Gortner, Jr. Dietary protein and the toxicity of sodium selenite in the white rat. J. Pharmacol. Exp. Ther. 68:292-299, 1940.
444. Liebscher, K., and H. Smith. Essential and nonessential trace elements. A method of determining whether an element is essential or nonessential in human tissue. Arch. Environ. Health 17:881-890, 1968.
445. Lindberg, P., and M. Sirén. Selenium concentration in kidneys of normal pigs and pigs affected with nutritional muscular dystrophy and liver dystrophy. Life Sci. 5:326-330, 1963.
446. Lisanti, L. E. Applicazione della fluorescenza da raggi X per la determinazione di alcuni costituenti minori nelle piante. Ricerca Sci. 28:2540-2544, 1958.
447. Lopez, P. L., R. L. Preston, and W. H. Pfander. *In vitro* uptake of selenium-75 by red blood cells from the immature ovine during varying selenium intakes. J. Nutr. 94:219-226, 1968.
448. Lopez, P. L., R. L. Preston, and W. H. Pfander. Whole-body retention, tissue distribution and excretion of selenium-75 after oral and intravenous administration in lambs fed varying selenium intakes. J. Nutr. 97:123-132, 1969.
449. Lott, P. F., P. Cukor, G. Moriber, and J. Solga. 2,3-Diaminonaphthalene as a reagent for the determination of milligram to submicrogram amounts of selenium. Anal. Chem. 35:1159-1163, 1963.
450. Lucy, J. A., A. T. Diplock, and C. P. Caygill. The intracellular distribution of acid-labile selenium in rat liver and the possible biological function of selenium and vitamin E. Biochem. J. 119:40P, 1970.

References

451. Ludwig, T. G., and B. G. Biddy. Geographic variations in the prevalence of dental caries in the United States of America. Caries Res. 3:32–43, 1969.
452. Luke, C. L. Photometric determination of traces of selenium (with diaminobenzidine) or tellurium in lead or copper. Anal. Chem. 31:572–574, 1959.
453. Luttrell, G. W. Annotated Bibliography on the Geology of Selenium. U.S. Geological Survey Bulletin No. 1019M. Washington, D.C.: U.S. Government Printing Office, 1959. 105 pp.
454. Maag, D. D., and M. W. Glenn. Toxicity of selenium: Farm animals, pp. 127–140. In O. H. Muth, Ed. Symposium: Selenium in Biomedicine. First International Symposium, Oregon State University, 1966. Westport, Conn.: AVI Publishing Co., 1967.
455. Maag, D. D., J. S. Osborn, and J. R. Clopton. The effect of sodium selenite on cattle. Amer. J. Vet. Res. 21:1049–1053, 1960.
456. Machado, E. A., E. A. Porta, W. S. Hartroft, and F. Hamilton. Studies on dietary hepatic necrosis. II. Ultrastructural and enzymatic alterations of the hepatocytic plasma membrane. Lab. Invest. 24:13–20, 1971.
457. Maguire, H. C., and A. M. Kligman. Lack of toxicity of selenium sulfide suspension for hair roots. J. Invest. Derm. 39:469–470, 1962.
458. Marjanen, H., and S. Soini. Possible causal relationship between nutritional imbalances, especially manganese deficiency and susceptibility to cancer, in Finland. Ann. Agric. Fenn. 11:391–406, 1972.
459. Martin, A. L. Toxicity of selenium to plants and animals. Amer. J. Botany 23:471–483, 1936.
460. Martin, A. L., and S. F. Trelease. Absorption of selenium by tobacco and soybeans in sand cultures. Amer. J. Botany 25:380–385, 1938.
461. Martin, J. L., and M. L. Gerlach. Separate elution by ion-exchange chromatography of some biologically important selenoamino acids. Anal. Biochem. 29:257–264, 1969.
462. Martin, J. L., A. Shrift, and M. L. Gerlach. Use of ^{75}Se-selenite from the study of selenium metabolism in *Astragalus*. Phytochemistry 10:945–952, 1971.
463. Mason, K. E., and J. O. Young. Effectiveness of selenium and zinc in protecting against cadmium-induced injury of the rat testis, pp. 383–394. In O. H. Muth, Ed. Symposium: Selenium in Biomedicine. First International Symposium, Oregon State University, 1966. Westport, Conn.: AVI Publishing Co., 1967.
464. Mason, K. E., J. O. Young, and J. E. Brown. Effectiveness of selenium and zinc in protection against cadmium-induced injury of the rat testis. Anat. Rec. 148:309, 1964.
465. Mathias, M. M., and D. E. Hogue. Effect of selenium, synthetic antioxidants, and vitamin E on the incidence of exudative diathesis in the chick. J. Nutr. 101:1399–1402, 1971.
466. Maxia, V., S. Meloni, M. A. Rollier, A. Brandone, V. N. Patwardhan, C. I. Waslien, and E. S. Shami. Selenium and chromium assay in Egyptian foods and in blood of Egyptian children by activation analysis, pp. 527–560. In Nuclear Activation Techniques in the Life Sciences 1972. Proceedings of a Symposium on Nuclear Activation Techniques in the Life Sciences Held by the International Atomic Energy Agency in Bled, Yugoslavia 10–14, April 1972. Vienna: International Atomic Energy Agency, 1972.
467. McConnell, K. P. Distribution and excretion studies in the rat after a single subtoxic subcutaneous injection of sodium selenate containing radioselenium. J. Biol. Chem. 141:427–437, 1941.
468. McConnell, K. P., and G. J. Cho. Active transport of L-selenomethionine in the intestine. Amer. J. Physiol. 213:150–156, 1967.

469. McConnell, K. P., and G. J. Cho. Transmucosal movement of selenium. Amer. J. Physiol. 208:1191–1195, 1965.
470. McConnell, K. P., and J. L. Hoffman. Methionine-selenomethionine parallels in *E. coli* polypeptide chain initiation and synthesis. Proc. Soc. Exp. Biol. Med. 140:638–641, 1972.
471. McConnell, K. P., and A. E. Kreamer. Incorporation of selenium-75 into dog hair. Proc. Soc. Exp. Biol. Med. 105:170–173, 1960.
472. McConnell, K. P., and R. S. Levy. Presence of selenium-75 in lipoproteins. Nature 195:774–776, 1962.
473. McConnell, K. P., and R. G. Martin. Biliary excretion of selenium in the dog after administration of sodium selenate containing radioselenium. J. Biol. Chem. 194:183–190, 1952.
474. McConnell, K. P., and O. W. Portman. Excretion of dimethyl selenide by the rat. J. Biol. Chem. 195:277–282, 1952.
475. McConnell, K. P., and O. W. Portman. Toxicity of dimethyl selenide in the rat and mouse. Proc. Soc. Exp. Biol. Med. 79:230–231, 1952.
476. McConnell, K. P., and D. M. Roth. Passage of selenium across the placenta and also into the milk of the dog. J. Nutr. 84:340–344, 1964.
477. McConnell, K. P., and D. M. Roth. ^{75}Se in rat intracellular liver fractions. Biochim. Biophys. Acta 62:503–508, 1962.
478. McConnell, K. P., and C. H. Wabnitz. Studies on the fixation of radioselenium in proteins. J. Biol. Chem. 226:765–776, 1957.
479. McConnell, K. P., C. H. Wabnitz, and D. M. Roth. Time-distribution studies of selenium-75 in dog serum proteins. Tex. Rep. Biol. Med. 18:438–445, 1960.
480. McCoy, K. E. M., and P. H. Weswig. Some selenium responses in the rat not related to vitamin E. J. Nutr. 98:383–389, 1969.
481. McCready, R. G. L., J. N. Campbell, and J. I. Payne. Selenite reduction by *Salmonella heidelberg*. Can. J. Microbiol. 12:703–714, 1966.
482. McNulty, J. S. Routine method for determining selenium in horticultural materials. Anal. Chem. 19:809–810, 1947.
483. Michel, R. L., C. K. Whitehair, and K. K. Keahey. Dietary hepatic necrosis associated with selenium–vitamin E deficiency in swine. J. Amer. Vet. Med. Assoc. 155:50–59, 1969.
484. Middleton, J. M. Selenium burn of the eye. Report of a case, with review of the literature. A.M.A. Arch. Ophthalmol. 38:806–811, 1947.
485. Millar, K. R. Distribution of Se^{75} in liver, kidney, and blood proteins of rats after intravenous injection of sodium selenite. N.Z. J. Agric. Res. 15:547–564, 1972.
486. Millar, K. R., and A. D. Sheppard. Vitamin E and selenium in cows' milk. N.Z. Med. J. 72:417, 1970.
487. Miller, D., J. H. Soares, Jr., P. Bauersfeld, Jr., and S. L. Cuppett. Comparative selenium retention by chicks fed sodium selenite, selenomethionine, fish meal and fish solubles. Poultry Sci. 51:1669–1673, 1972.
488. Miller, J. T., and H. G. Byers. A selenium spring. Ind. Eng. Chem. News Ed. 13:456, 1935.
489. Miller, J. T., and H. G. Byers. Selenium in plants in relation to its occurrence in soils. J. Agric. Res. 55:59–68, 1937.
490. Miller, W. T., and H. W. Schoening. Toxicity of selenium fed to swine in the form of sodium selenite. J. Agric. Res. 56:831–842, 1938.
491. Miller, W. T., and K. T. Williams. Effect of feeding repeated small doses of selenium as sodium selenite to equines. J. Agric. Res. 61:353–368, 1940.

References

492. Miller, W. T., and K. T. Williams. Minimum lethal doses of selenium, as sodium selenite, for horses, mules, cattle, and swine. J. Agric. Res. 60:163–173, 1940.
493. Minyard, J. A. Selenium poisoning in beef cattle. S. Dakota Farm Home Res. 12:1–2, 1961.
494. Mondragon, M. C. de, and W. G. Jaffé. Selenio en alimentos y en orina de escolares de diferentes zonas de Venezuela. Arch. Latinoamer. Nutr. 21:185–195, 1971.
495. Money, D. F. L. Cot deaths and deficiency of vitamin E and selenium. Brit. Med. J. 4:559, 1971.
496. Money, D. F. L. Vitamin E and selenium deficiencies and their possible aetiological role in the sudden death in infants syndrome. N.Z. Med. J. 71:32–34, 1970.
497. Money, D. F. L. Vitamin E and selenium deficiencies and their possible etiological role in the sudden death in infants syndrome. J. Pediat. 77:165–166, 1970.
498. Morette, A., and J. P. Diven. La détermination du sélénium dans l'eau. Ann. Pharm. Franc. 23:169–178, 1965.
499. Morris, V. C., and O. A. Levander. Selenium content of foods. J. Nutr. 100:1383–1388, 1970.
500. Morrow, D. A. Acute selenite toxicosis in lambs. J. Amer. Vet. Med. Assoc. 152:1625–1629, 1968.
501. Morss, S. G., H. R. Ralston, and H. S. Olcott. Selenium determination in human serum lipoprotein fractions by neutron activation analysis. Anal. Biochem. 49:598–601, 1972.
502. Motley, H. L., M. M. Ellis, and M. D. Ellis. Acute sore throats following exposure to selenium. J.A.M.A. 109:1718–1719, 1937.
503. Moxon, A. L. Alkali Disease or Selenium Poisoning. South Dakota Agricultural Experiment Station Bulletin No. 311. Brookings: South Dakota State College of Agriculture and Mechanic Arts, Agricultural Experiment Station, 1937. 91 pp.
504. Moxon, A. L. The influence of arsenic on selenium poisoning in hogs. Proc. S. Dakota Acad. Sci. 21:34–36, 1941.
505. Moxon, A. L. Some Factors Influencing the Toxicity of Selenium. Ph.D. thesis, Madison: University of Wisconsin, 1941.
506. Moxon, A. L. The effect of arsenic on the toxicity of seleniferous grains. Science 88:81, 1938.
507. Moxon, A. L. Toxicity of selenium-cystine and some other organic selenium compounds. J. Amer. Pharm. Assoc. Sci. Ed. 29:249–251, 1940.
508. Moxon, A. L., H. D. Anderson, and E. P. Painter. The toxicity of some organic selenium compounds. J. Pharmacol. Exp. Ther. 63:357–368, 1938.
509. Moxon, A. O., O. E. Olson, and W. V. Searight. Selenium in Rocks, Soils and Plants. South Dakota Agricultural Experiment Station Technical Bulletin No. 2. Brookings: South Dakota State College of Agriculture and Mechanic Arts, Agricultural Experiment Station, 1939. 94 pp.
510. Moxon, A. L., O. E. Olson, E. I. Whitehead, R. J. Hilmoe, and S. N. White. Selenium distribution in milled seleniferous wheats. Cereal Chem. 20:376–380, 1943.
511. Moxon, A. L., and W. E. Poley. The relation of selenium content of grains in the ration to the selenium content of poultry carcass and eggs. Poultry Sci. 17:77–80, 1938.
512. Moxon, A. L., and M. Rhian. Loss of selenium by various grains during storage. Proc. S. Dakota Acad. Sci. 18:20–22, 1938.
513. Moxon, A. L., and M. A. Rhian. Selenium poisoning. Physiol. Rev. 23:305–337, 1943.
514. Moxon, A. L., M. A. Rhian, H. D. Anderson, and O. E. Olson. Growth of steers on seleniferous range. J. Animal Sci. 2:299–309, 1944.
515. Mudd, S. H., and G. L. Cantoni. Selenomethionine in enzymatic transmethylations. Nature 180:1052, 1957.

516. Mühlemann, H. R., and K. G. König. The effect of some trace elements on experimental fissure caries and on growth in Osborne–Mendel rats. Helv. Odontol. Acta 8(Suppl.):79–81, 1968.
517. Muhler, J. C., and W. G. Shafer. The effect of selenium on the incidence of dental caries in rats. J. Dent. Res. 36:895–896, 1957.
518. Munsell, H. E., G. M. DeVaney, and M. H. Kennedy. Toxicity of Food Containing Selenium as Shown by Its Effect on the Rat. U.S. Department of Agriculture Technical Bulletin No. 534. Washington, D.C.: U.S. Department of Agriculture, 1936. 25 pp.
519. Muth, O. H. [Discussion of clinical aspects of selenium deficiency], p. 225. In O. H. Muth, Ed. Symposium: Selenium in Biomedicine. First International Symposium, Oregon State University, 1966. Westport, Conn.: AVI Publishing Co., 1967.
520. Muth, O. H. Selenium-responsive disease of sheep. J. Amer. Vet. Med. Assoc. 157:1507–1511, 1970.
521. Muth, O. H., Ed. Symposium: Selenium in Biomedicine. First International Symposium, Oregon State University, 1966. Westport, Conn.: AVI Publishing Co., 1967. 445 pp.
522. Muth, O. H. Theme of the symposium: The biomedical aspects of selenium, pp. 3–6. In O. H. Muth, Ed. Symposium: Selenium in Biomedicine. First International Symposium, Oregon State University, 1966. Westport, Conn.: AVI Publishing Co., 1967.
523. Muth, O. H. White muscle disease, a selenium-responsive myopathy. J. Amer. Vet. Med. Assoc. 142:272–277, 1963.
524. Muth, O. H. White muscle disease (myopathy) in lambs and calves. I. Occurrence and nature of the disease under Oregon conditions. J. Amer. Vet. Med. Assoc. 126:355–361, 1955.
525. Muth, O. H., and W. Binns. Selenium toxicity in domestic animals. Ann. N.Y. Acad. Sci. 111:583–590, 1964.
526. Muth, O. H., J. E. Oldfield, L. F. Remmert, and J. R. Schubert. Effects of selenium and vitamin E on white muscle disease. Science 128:1090, 1958.
527. Muth, O. H., J. E. Oldfield, J. R. Schubert, and L. F. Remmert. White muscle disease (myopathy) in lambs and calves. VI. Effects of selenium and vitamin E on lambs. Amer. J. Vet. Res. 20:231–234, 1959.
528. Muth, O. H., P. H. Weswig, P. D. Whanger, and J. E. Oldfield. Effect of feeding selenium-deficient ration to the subhuman primate (*Saimiri sciureus*). Amer. J. Vet. Res. 32:1603–1605, 1971.
529. Muth, O. H., P. D. Whanger, P. H. Weswig, and J. E. Oldfield. Occurrence of myopathy in lambs of ewes fed added arsenic in a selenium-deficient ration. Amer. J. Vet. Res. 32:1621–1623, 1971.
530. Nadkarni, R. A., and W. D. Ehmann. Instrumental neutron activation analysis of tobacco products, pp. 190–196. In J. R. DeVoe, Ed. Modern Trends in Activation Analysis. Proceedings of 1968 International Conference Held at National Bureau of Standards, Gaithersburg, Maryland, October 7–11, 1968. National Bureau of Standards Special Publication No. 312. Washington, D.C.: U.S. Government Printing Office, 1969.
531. Nadkarni, R. A., and W. D. Ehmann. Neutron activation analysis of wheat flour samples. Radiochem. Radioanal. Lett. 6:89–96, 1971.
532. Nagai, I. An experimental study of selenium poisoning. Igaku Kenkyu (Acta Medica) 29:1505–1532, 1959. (in Japanese; summary in English)
533. Naidu, P. P., and G. G. Rao. Colorimetric determination of traces of selenium, tellurium, manganese, and cerium. Talanta 17:817–822, 1970.

References

534. Nakashima, S., and K. Toei. Determination of ultramicro amounts of selenium by gas chromatography. Talanta 15:1475-1476, 1968.
535. National Research Council, Agricultural Board, Committee on Animal Nutrition, Subcommittee on Selenium. Selenium in Nutrition. Washington, D.C.: National Academy of Sciences, 1971. 79 pp.
536. Natusch, D. F. S., J. R. Wallace, and C. A. Evans, Jr. Toxic trace elements: Preferential concentration in respirable particles. Science 183:202-204, 1974.
537. Navia, J. M., L. Menaker, J. Seltzer, and R. S. Harris. Effect of Na_2SeO_3 supplemented in the diet or water on dental caries of rats. Fed. Proc. 27:676, 1968. (abstract)
538. Neethling, L. P., J. M. M. Brown, and P. J. DeWet. Natural occurrence of selenium in sheep blood and tissues and its possible biological effects. J. S. Afr. Vet. Med. Assoc. 39(1):93-97, 1968.
539. Neethling, L. P., J. M. M. Brown, and P. J. DeWet. The toxicology and metabolic fate of selenium in sheep. J. S. Afr. Vet. Med. Assoc. 39(3):25-33, 1968.
540. Nelson, A. A., O. G. Fitzhugh, and H. O. Calvery. Liver tumors following cirrhosis caused by selenium in rats. Cancer Res. 3:230-236, 1943.
541. Nesheim, M. C., and M. L. Scott. Nutritional effects of selenium compounds in chicks and turkeys. Fed. Proc. 20:674-678, 1961.
542. Nesheim, M. C., and M. L. Scott. Studies on the nutritive effects of selenium in chicks. J. Nutr. 65:601-618, 1958.
543. Neubert, D., and A. L. Lehninger. The effect of thiols and disulfides on water uptake and extrusion by rat liver mitochondria. J. Biol. Chem. 237:952-958, 1962.
544. Newberry, C. L., and G. D. Christian. Rapid colorimetric determination of microgram amounts of selenium in soils and biological samples. J. Assoc. Off. Agric. Chemists 48:322-326, 1965.
545. Nickerson, W. J., and G. Falcone. Enzymatic reduction of selenite. J. Bacteriol. 85:763-771, 1963.
546. Nigam, S. N., and W. B. McConnell. Isolation and identification of L-cystathionine and L-selenocystathionine from the foliage of *Astragalus pectinatus*. Phytochemistry 11:377-380, 1972.
547. Nigam, S. N., and W. B. McConnell. Selenium amino compounds in *Astragalus pectinatus:* Isolation and identification of two glutamyl selenocystathionines from the seeds. Fed. Proc. 32:886, 1973. (abstract)
548. Nigam, S. N., and W. B. McConnell. Seleno amino compounds from *Astragalus bisulcatus* isolation and identification of γ-L-glutamyl-Se-methylseleno-L-cysteine and Se-methylseleno-L-cysteine. Biochim. Biophys. Acta 192:185-190, 1969.
549. Nigam, S. N., J. I. Tu, and W. B. McConnell. Distribution of selenomethylselenocysteine and some other amino acid in species of *Astragalus*, with special reference of their distribution during the growth of *A. bisulcatus*. Phytochemistry 8:1161-1165, 1969.
550. Nissen, P., and A. A. Benson. Absence of selenate esters and "selenolipid" in plants. Biochim. Biophys. Acta 82:400-402, 1964.
551. Nixon, G. S., and V. B. Myers. Estimation of selenium in human dental enamel by activation analysis. Caries Res. 4:179-187, 1970.
552. Noguchi, T., A. H. Cantor, and M. L. Scott. Mode of action of selenium and Vitamin E in prevention of exudative diathesis in chicks. J. Nutr. 103:1502-1511, 1973.
553. Obermeyer, B. D., I. S. Palmer, O. E. Olson, and A. W. Halverson. Toxicity of trimethylselenonium chloride in the rat with and without arsenite. Toxicol. Appl. Pharmacol. 20:135-146, 1971.
554. Ochoa-Solano, A., and C. Gitler. Incorporation of ^{75}Se-selenomethionine and ^{35}S-methionine into chicken egg white proteins. J. Nutr. 94:243-248, 1968.

555. Oelschläger, W. Selen in der Ernährung. Umschau 70:45, 1970.
556. Oelschläger, W., and K. H. Menke. Über Selengehalte pflanzlicher, tierischer und anderer Stoffe. 2. Mitteilung: Selen- und Schwefelgehalte in Nahrungsmitteln. Z. Ernährungswissenschaft 9:216-222, 1969.
557. Okamota, Y., and W. H. H. Günther, Eds. Organic selenium and tellurium chemistry. Ann. N.Y. Acad. Sci. 192:1-226, 1972.
558. Oksanen, H. E. Selenium deficiency: Clinical aspects and physiological responses in farm animals, pp. 215-229. In O. H. Muth, Ed. Symposium: Selenium in Biomedicine. First International Symposium, Oregon State University, 1966. Westport, Conn.: AVI Publishing Co., 1967.
559. Olson, E. C., and J. W. Shell. The simultaneous determination of traces of selenium and mercury in organic compounds by x-ray fluorescence. Anal. Chim. Acta 23:219-224, 1960.
560. Olson, O. E. Fluorometric analysis of selenium in plants. J. Assoc. Off. Anal. Chemists 52:627-634, 1969.
561. Olson, O. E. Selenium as a toxic factor in animal nutrition, pp. 68-78. In Proceedings of the Georgia Nutrition Conference, Atlanta, 1969.
562. Olson, O. E. Selenium in feedstuffs: Deficiencies and excesses, pp. 7-13. In Proceedings of the Thirty-first Annual Minnesota Nutrition Conference, 1970.
563. Olson, O. E. Soil, plant, animal cycling of excessive levels of selenium, pp. 297-312. In O. H. Muth, Ed. Symposium: Selenium in Biomedicine. First International Symposium, Oregon State University, 1966. Westport, Conn.: AVI Publishing Co., 1967.
564. Olson, O. E. The absorption of selenium by certain inorganic colloids. Proc. S. Dakota Acad. Sci. 19:22-24, 1939.
565. Olson, O. E., C. A. Dinkel, and L. D. Kamstra. A new aid in diagnosing selenium poisoning. S. Dakota Farm Home Res. 6:12-14, 1954.
566. Olson, O. E., and L. B. Embry. Chronic selenite toxicity in cattle. Proc. S. Dakota Acad. Sci. 52:50-58, 1973.
567. Olson, O. E., and A. W. Halverson. Effect of linseed oil meal and arsenicals on selenium poisoning in the rat. Proc. S. Dakota Acad. Sci. 33:90-94, 1954.
568. Olson, O. E., and C. W. Jensen. The absorption of selenate and selenite selenium by colloidal ferric hydroxide. Proc. S. Dakota Acad. Sci. 20:115-121, 1940.
569. Olson, O. E., D. F. Jornlin, and A. L. Moxon. The selenium content of vegetation and the mapping of seleniferous soils. J. Amer. Soc. Agron. 34:607-615, 1942.
570. Olson, O. E., E. J. Navacek, E. I. Whitehead, and I. S. Palmer. Investigations on selenium in wheat. Phytochemistry 9:1181-1188, 1970.
571. Olson, O. E., I. S. Palmer, and E. I. Whitehead. Determination of selenium in biological materials, pp. 39-78. In D. Glick, Ed. Methods of Biochemical Analysis. Volume 21. New York: John Wiley & Sons, 1973.
572. Olson, O. E., B. M. Schulte, E. I. Whitehead, and A. W. Halverson. Effect of arsenic on selenium metabolism in rats. J. Agric. Food Chem. 11:531-534, 1963.
573. Olson, O. E., E. I. Whitehead, and A. L. Moxon. Occurrence of soluble selenium in soils and its availability to plants. Soil Sci. 54:47-53, 1942.
574. Orentreich, N., E. H. Taylor, R. A. Berger, and R. Auerbach. Comparative study of two antidandruff preparations. J. Pharm. Sci. 58:1279-1280, 1969.
575. Orstadius, K. Toxicity of a single subcutaneous dose of sodium selenite in pigs. Nature 188:1117, 1960.
576. Orstadius, K., G. Nordström, and N. Lannek. Combined therapy with vitamin E and selenite in experimental nutritional muscular dystrophy of pigs. Cornell Vet. 53:60-71, 1963.

References

577. Osburn, R. L., A. D. Shendrikar, and P. W. West. New spectrophotometric method for determination of submicrogram quantities of selenium. Anal. Chem. 43:594–597, 1971.
578. Painter, E. P. The chemistry and toxicity of selenium compounds with special reference to the selenium problem. Chem. Rev. 28:179–213, 1941.
579. Painter, E. P., and K. W. Franke. On the relationship of selenium to sulfur and nitrogen deposition in cereals. Amer. J. Botany 27:336–339, 1940.
580. Painter, E. P., and K. W. Franke. Selenium in proteins from toxic foodstuffs II. The effects of acid hydrolysis. Cereal Chem. 13:172–178, 1936.
581. Painter, E. P., and K. W. Franke. Selenium in proteins from toxic foodstuffs. III. The removal of selenium from toxic protein hydrolysates. J. Biol. Chem. 111:643–651, 1935.
582. Painter, E. P., and K. W. Franke. The decomposition of seleniferous proteins in alkaline solutions. J. Biol. Chem. 134:557–566, 1940.
583. Pakkala, I. S., W. H. Gutenmann, D. J. Lisk, G. E. Burdick, and E. J. Harris. A survey of the selenium content of fish from 49 New York State waters. Pest. Monitor. J. 6:107–114, 1972.
584. Palmer, I. S., D. D. Fischer, A. W. Halverson, and O. E. Olson. Identification of a major selenium excretory product in rat urine. Biochim. Biophys. Acta 177:336–342, 1969.
585. Palmer, I. S., R. P. Gunsalus, A. W. Halverson, and O. E. Olson. Trimethyl-selenonium ion as a general excretory product from selenium metabolism in the rat. Biochim. Biophys. Acta 208:260–266, 1970.
586. Palmer, I. S., and O. E. Olson. Relative toxicities of selenite and selenate in the drinking water of rats. J. Nutr. 104:306–314, 1974.
587. Pařízek, J. The destructive effect of cadmium on testicular tissue and its prevention by zinc. J. Endocrinol. 15:56–63, 1957.
588. Pařízek, J. Vascular changes at sites of oestrogen biosynthesis produced by parenteral injection of cadmium salts: The destruction on placenta by cadmium salts. J. Reprod. Fertil. 7:263–265, 1964.
589. Pařízek, J., I. Ošťádalová, J. Kaloušková, A. Babický, and J. Beneš. The detoxifying effects of selenium interrelations between compounds of selenium and certain metals, pp. 85–122. In W. Mertz and W. E. Cornatzer, Eds. Newer Trace Elements in Nutrition. New York: Marcel Dekker, Inc., 1971.
590. Pařízek, J., I. Ošťádalová, J. Kaloušková, A. Babický, L. Pavlík, and B. Bíbr. Effect of mercuric compounds on the maternal transmission of selenium in the pregnant and lactating rat. J. Reprod. Fertil. 25:157–170, 1971.
591. Parker, C. A., and L. G. Harvey. Fluorimetric determination of sub-microgram amounts of selenium. Analyst 86:54–62, 1961.
592. Patrias, G., and O. E. Olson. Selenium contents of samples of corn from midwestern states. Feedstuffs 41(43):32–34, 1969.
593. Patrick, H., R. A. Voitle, H. M. Hyre, and W. G. Martin. Incorporation of phosphorus^{32} and selenium^{75} in cock sperm. Poultry Sci. 44:587–591, 1965.
594. Patterson, E. L., R. Milstrey, and E. L. R. Stokstad. Effect of selenium in preventing exudative diathesis in chicks. Proc. Soc. Exp. Biol. Med. 95:617–620, 1957.
595. Paulson, G. D., C. A. Baumann, and A. L. Pope. Fate of a physiological dose of selenate in the lactating ewe: Effect of sulfate. J. Animal Sci. 25:1054–1058, 1966.
596. Paulson, G. D., C. A. Baumann, and A. L. Pope. Metabolism of ^{75}Se-selenite, ^{75}Se-selenate, ^{75}Se-selenomethionine and ^{35}S-sulfate by rumen microorganisms in vitro. J. Animal Sci. 27:497–504, 1968.
597. Paulson, G. D., G. A. Broderick, C. A. Baumann, and A. L. Pope. Effect of feeding

sheep selenium fortified trace mineralized salt. Effect of tocopherol. J. Animal Sci. 27:195–202, 1968.
598. Pearsall, W. H. The soil complex in relation to plant communities. I. Oxidation-reduction potentials in soils. J. Ecology 26:180–193, 1938.
599. Pellegrini, L. A Study of Vitamin E Deficiency in Pigs Fed a Torula Yeast Diet. Ph.D. thesis, St. Paul: University of Minnesota, 1958. 105 pp.
600. Pendell, H. W., O. H. Muth, J. E. Oldfield, and P. H. Weswig. Tissue fatty acids in normal and myopathic lambs. J. Animal Sci. 29:94–98, 1969.
601. Penington, D. G. Assessment of platelet production with ^{75}Se selenomethionine. Brit. Med. J. 4:782–784, 1969.
602. Perkins, A. T., and H. H. King. Selenium and Tenmarq wheat. J. Amer. Soc. Agron. 30:664–667, 1938.
603. Peterson, P. J., and G. W. Butler. Colloidal selenium availability to three pasture species in pot culture. Nature 212:961–962, 1966.
604. Peterson, P. J., and G. W. Butler. Significance of selenocystathionine in an Australian selenium-accumulating plant, *Neptunia amplexicaulis*. Nature 213:599–600, 1967.
605. Peterson, P. J., and G. W. Butler. The uptake and assimilation of selenite by higher plants. Austral. J. Biol. Sci. 15:126–246, 1962.
606. Peterson, P. J., and D. J. Spedding. The excretion by sheep of ^{75}selenium into red clover (*Trifolium pratense* L.): The chemical nature of the excreted selenium and its uptake by three plant species. N.Z. J. Agric. Res. 6:13–23, 1963.
607. Pillay, K. K. S., and C. C. Thomas, Jr. Determination of the trace element levels in atmospheric pollutants by neutron activation analysis. J. Radioanal. Chem. 7: 107–118, 1971.
608. Pillay, K. K. S., C. C. Thomas, Jr., and J. W. Kaminski. Neutron activation analysis of the selenium content of fossil fuels. Nucl. Appl. Technol. 7:478–483, 1969.
609. Pillay, K. K. S., C. C. Thomas, Jr., and J. A. Sondel. Activation analysis of airborne selenium as a possible indicator of atmospheric sulfur pollutants. Environ. Sci. Technol. 5:74–77, 1971.
610. Pinsent, J. The need for selenite and molybdate in the formation of formic dehydrogenase by members of the *Coli aerogenes* group of bacteria. Biochem. J. 57:10–16, 1954.
611. Pletnikova, I. P. Biological effect and safe concentration of selenium in drinking water. Hyg. Sanit. 35(1–3):176–181, 1970.
612. Poley, W. E., W. O. Wilson, A. L. Moxon, and J. B. Taylor. The effect of selenized grains on the rate of growth in chicks. Poultry Sci. 20:171–179, 1941.
613. Pond, W. G., W. H. Allaway, E. F. Walker, Jr., and L. Krook. Effects of corn selenium content and drying temperature and of supplemental vitamin E on growth, liver selenium and blood vitamin E content of chicks. J. Animal Sci. 33:996–1000, 1971.
614. Potchen, E. J. Isotopic labeling of the rat parathyroid as demonstrated by autoradiography. J. Nucl. Med. 4:480–484, 1963.
615. Pringle, P. Occupational dermatitis following exposure to inorganic selenium compounds. Brit. J. Derm. Syphil. 54:54–58, 1942.
616. Rahman, M. M., R. E. Davies, C. W. Deyoe, and J. R. Couch. Selenium and torula yeast in production of exudative diathesis in chicks. Proc. Soc. Exp. Biol. Med. 105:227–230, 1960.
617. Rancitelli, L. A., J. A. Cooper, and R. W. Perkins. The multielement analysis of biological material by neutron activation and direct instrumental techniques, pp. 101–109. In J. R. DeVoe, Ed. Modern Trends in Activation Analysis. Proceedings of 1968 International Conference Held at National Bureau of Standards,

Gaithersburg, Maryland, October 7–11, 1968. Vol. I. National Bureau of Standards Special Publication No. 312. Washington, D.C.: U.S. Government Printing Office, 1969.
618. Rancitelli, L. A., R. E. Perkins, T. M. Tanner, and C. W. Thomas. Stable elements of the atmosphere as tracers of precipitation scavenging, pp. 99–108. In Precipitation Scavenging. Proceedings of a Symposium Held at Richland, Washington, June 2–4, 1970. Sponsored by Pacific Northwest Laboratory, Battelle Memorial Institute, and Fallout Studies Branch, Division of Biology and Medicine, U.S. Atomic Energy Commission. Springfield, Va.: National Technical Information Service, 1970.
619. Ransone, J. W., N. M. Scott, Jr., and E. C. Knoblock. Selenium sulfide intoxication. New Engl. J. Med. 264:384–385, 1961.
620. Ravikovitch, S., and M. Margolin. Selenium in soils and plants. Ktavim. Rec. Agric. Res. Stat. (Rehovot, Israel) 7(2–3):41–52, 1957.
621. Recommendations of Permissible Criteria of Hazardous Working Environments. Japanese Association of Industrial Health, 1969. 8 pp.
622. Revici, E. The control of cancer with lipids. Clin. Path. Conf. Beth David Hospital, New York, May 1955.
623. Rhead, W. J., E. E. Cary, W. H. Allaway, S. L. Saltzstein, and G. N. Schrauzer. The vitamin E and selenium status of infants and the sudden infant death syndrome. Bioinorgan. Chem. 1:289–294, 1972.
624. Rhead, W. J., G. N. Schrauzer, S. L. Saltzstein, E. E. Cary, and W. H. Allaway. Vitamin E, selenium, and the sudden infant death syndrome. J. Pediat. 81:415–416, 1972.
625. Rhian, M., and A. L. Moxon. Chronic selenium poisoning in dogs and its prevention by arsenic. J. Pharmacol. Exp. Ther. 78:249–264, 1943.
626. Robbins, C. W., and D. L. Carter. Selenium concentrations in phosphorus fertilizer materials and associated uptake by plants. Soil Sci. Soc. Amer. Proc. 34:506–509, 1970.
627. Roberts, M. E. Antiinflammation studies. II. Antiinflammatory properties of selenium. Toxicol. Appl. Pharmacol. 5:500–506, 1963.
628. Robertson, D. S. F. Selenium, a possible teratogen? Lancet 1:518, 1970.
629. Robinson, H. M., Jr., and S. N. Yaffe. Selenium sulfide in the treatment of pityriasis versicolor. J.A.M.A. 162:113–114, 1956.
630. Robinson, W. O. Selenium content of wheat from various parts of the world. Ind. Eng. Chem. Ind. Ed. 28:736–738, 1936.
631. Robinson, W. O., H. C. Dudley, K. T. Williams, and H. G. Byers. Determination of selenium and arsenic by distillation in pyrites, shales, soils and agricultural products. Ind. Eng. Chem. Anal. Ed. 6:274–276, 1934.
632. Rodriguez-Antunez, A., T. A. Egleston, E. J. Fillson, B. H. Sullivan, and C. H. Brown. Pancreatic scanning: Cleveland Clinic experience. Lahey Clin. Bull. 17:29–33, 1968.
633. Roffler, S. A., T. F. Allsop, and E. W. Wright. Selenium transport and plasma lipoproteins. N.Z. J. Sci. 11:258–263, 1968.
634. Rook, H. L. Rapid, quantitative separation for the determination of selenium using neutron activation analysis. Anal. Chem. 44:1276–1278, 1972.
635. Rosenfeld, I. Metabolic Effects and Metabolism of Selenium in Animals. Wyoming Agricultural Experiment Station Bulletin No. 414. Laramie: University of Wyoming, Agricultural Experiment Station, 1964. 64 pp.
636. Rosenfeld, I., and O. A. Beath. Chemistry of selenium, pp. 299–332. In Selenium. Geobotany, Biochemistry, Toxicity, and Nutrition. New York: Academic Press, 1964.
637. Rosenfeld, I., and O. A. Beath. Congenital malformation of eyes of sheep. J. Agric. Res. 75:93–103, 1947.

638. Rosenfield, I., and O. A. Beath. Effect of selenium on reproduction in rats. Proc. Soc. Exp. Biol. Med. 87:295-297, 1954.
639. Rosenfeld, I., and O. A. Beath. Metabolism of sodium selenite by the tissues. J. Biol. Chem. 172:333-341, 1948.
640. Rosenfeld, I., and O. A. Beath. Pathology of Selenium Poisoning. Wyoming Agricultural Experiment Station Bulletin No. 275. Laramie: University of Wyoming, Agricultural Experiment Station, 1946. 27 pp.
641. Rosenfeld, I., and O. A. Beath. Selenium. Geobotany, Biochemistry, Toxicity, and Nutrition. New York: Academic Press, 1964. 411 pp.
642. Rosenfeld, I., and F. P. Wirtz. Biochemical and chemical studies on *Astragalus* leaves and roots, pp. 1-42. In Wyoming Agricultural Experiment Station Bulletin No. 385. Laramie: University of Wyoming Agricultural Experiment Station, 1962.
643. Rotruck, J. T., W. G. Hoekstra, and A. L. Pope. Glucose-dependent protection by dietary selenium against haemolysis of rat erythrocytes *in vitro*. Nature New Biol. 231:223-224, 1971.
644. Rotruck, J. T., A. L. Pope, H. E. Ganther, and W. G. Hoekstra. Prevention of oxidative damage to rat erythrocytes by dietary selenium. J. Nutr. 102:689-696, 1972.
645. Rotruck, J. T., A. L. Pope, H. E. Ganther, A. B. Swanson, D. G. Hafeman, and W. G. Hoekstra. Selenium: Biochemical role as a component of glutathione. Science 179:588-590, 1973.
646. Ruth, G. R., and J. F. van Vleet. Experimentally induced selenium-vitamin E deficiency in growing swine: Selective destruction of type I skeletal muscle fibers. Amer. J. Vet. Res. 35:237-244, 1974.
647. Salisbury, R. M., J. Edmondson, W. S. H. Poole, F. C. Bobby, and H. Birnie. Exudative diathesis and white muscle disease of poultry in New Zealand, pp. 379-384. In Proceedings of the Twelfth World's Poultry Congress, Sydney, Australia, 1962.
648. Sauer, G. C. Treatment of seborrheic dermatitis ("dandruff") with selenium sulfide suspension. J. Missouri State Med. Assoc. 49:911-912, 1952.
649. Sawicki, E. New color test for selenium. Anal. Chem. 29:1376-1377, 1957.
650. Scala, J., and H. H. Williams. The enhancement of selenite toxicity by methionine in *Escherichia coli*. Arch. Biochem. Biophys. 99:363-368, 1962.
651. Scarabicchi, S., M. Merlini, and D. Ribaldone. Modificazioni elettrocardiografiche indotte sperimentalmente da sali inorganici di selenio. Minerva Dietol. 8:222-227, 1968.
652. Schoental, R. Selenium-75 in the Harderian glands and brown fat of rats given sodium selenite labelled with selenium-75. Nature 218:294-295, 1968.
653. Schrauzer, G. N., and W. J. Rhead. Interpretation of the methylene blue reduction test of human plasma and the possible cancer protection effect of selenium. Experientia 27:1069-1071, 1971.
654. Schroeder, H. A. Cadmium as a factor in hypertension. J. Chronic Dis. 18:647-656, 1965.
655. Schroeder, H. A. Effects of selenate, selenite, and tellurite on the growth and early survival of mice and rats. J. Nutr. 92:334-338, 1967.
656. Schroeder, H. A., D. V. Frost, and J. J. Balassa. Essential trace metals in man: Selenium. J. Chronic Dis. 23:227-243, 1970.
657. Schroeder, H. A., and M. Mitchener. Selenium and tellurium in mice. Effects on growth, survival, and tumors. Arch. Environ. Health 24:66-71, 1972.
658. Schroeder, H. A., and M. Mitchener. Selenium and tellurium in rats: Effect on growth, survival, and tumors. J. Nutr. 101:1531-1540, 1971.

659. Schroeder, H. A., and M. Mitchener. Toxic effects of trace elements on reproduction of mice and rats. Arch. Environ. Health 23:102–106, 1971.
660. Schubert, J. R., O. H. Muth, J. E. Oldfield, and L. F. Remmert. Experimental results with selenium in white muscle disease of lambs and calves. Fed. Proc. 20:689–694, 1961.
661. Schultz, J., and H. E. Lewis. The excretion of volatile selenium compounds after the administration of sodium selenite to white rats. J. Biol. Chem. 133:199–207, 1940.
662. Schumann, H., and W. Kölling. Eine vereinfachte Methode zur Bestimmung von Selen in Schlammen und Stauben der Schwefelsäureindustrie. Z. Chem. 1:371–372, 1961.
663. Schwarz, K. A hitherto unrecognized factor against dietary necrotic liver degeneration in American yeast (Factor 3). Proc. Soc. Exp. Biol. Med. 78:852–856, 1951.
664. Schwarz, K. [Discussion at end of chapter on selenium deficiency], pp. 225–226. In O. H. Muth, Ed. Symposium: Selenium in Biomedicine. First International Symposium, Oregon State University, 1966. Westport, Conn.: AVI Publishing Co., 1967.
665. Schwarz, K., and C. N. Foltz. Factor 3 activity of selenium compounds. J. Biol. Chem. 233:245–251, 1958.
666. Schwarz, K., and C. M. Foltz. Selenium as an integral part of Factor 3 against dietary necrotic liver degeneration. J. Amer. Chem. Soc. 79:3292–3293, 1957.
667. Schwarz, K., L. A. Porter, and A. Fredga. Some regularities in the structure–function relationship of organoselenium compounds effective against dietary liver necrosis. Ann. N.Y. Acad. Sci. 192:200–214, 1972.
668. Schwarz, K., J. A. Stesney, and C. N. Foltz. Relation between selenium traces in L-cystine and protection against dietary liver necrosis. Metab. Clin. Exp. 8:88–90, 1959.
669. Schwarz, K., and E. Sweeney. Selenite binding to sulfur amino acids. Fed. Proc. 23:421, 1964.
670. Scott, M. L., and A. H. Cantor. Tissue selenium levels in chicks receiving graded amounts of dietary selenium. Fed. Proc. 30:237, 1971.
671. Scott, M. L., G. Olson, L. Krook, and W. R. Brown. Selenium-responsive myopathies of myocardium and of smooth muscle in the young poult. J. Nutr. 91:573–583, 1967.
672. Scott, M. L., and J. N. Thompson. Selenium content of feedstuffs and effects of dietary selenium levels upon tissue selenium in chicks and poults. Poultry Sci. 50:1742–1748, 1971.
673. Selenium and cancer. Nutr. Rev. 28:75–80, 1970.
674. Selenium poisons. Indians. Sci. News Lett. 81:254, 1962.
675. Sellers, E. A., R. W. You, and C. C. Lucas. Lipotropic agents in liver damage produced by selenium or carbon tetrachloride. Proc. Soc. Exp. Biol. Med. 75:118–121, 1950.
676. Selyankina, K. P. Selenium and tellurium in the atmosphere around electrolytic copper plants. Hyg. Sanit. 35(3):431–432, 1970.
677. Shah, K. R., R. H. Filby, and A. I. Davis. Determination of trace elements in tea and coffee by neutron activation analysis. Int. J. Environ. Anal. Chem. 1:63–73, 1971.
678. Shamberger, R. J. Is selenium a teratogen? Lancet 2:1316, 1971.
679. Shamberger, R. J. Relation of selenium to cancer. I. Inhibitory effect of selenium on carcinogenesis. J. Nat. Cancer Inst. 44:931–936, 1970.
680. Shamberger, R. J., and D. V. Frost. Possible protective effect of selenium against human cancer. Can. Med. Assoc. J. 100:682, 1969.

681. Shamberger, R. J., and G. Rudolph. Protection against cocarcinogenesis by antioxidants. Experientia 22:116, 1966.
682. Shamberger, R. J., E. Rukovena, A. K. Longfield, S. A. Tytko, S. Deodhar, and C. E. Willis. Antioxidants and cancer. I. Selenium in the blood of normals and cancer patients. J. Nat. Cancer Inst. 50:863-870, 1973.
683. Shamberger, R. J., S. Tytko, and C. E. Willis. Antioxidants in cereals and in food preservatives and declining gastric cancer mortality. Cleveland Clin. Quart. 39: 119-124, 1972.
684. Shamberger, R. J., and C. E. Willis. Selenium distribution and human cancer mortality. Crit. Rev. Clin. Lab. Sci. 2:211-221, 1971.
685. Shearer, T. R. The postdevelopmental and developmental uptake of dietary selenomethionine by teeth. Fed. Proc. 33:694, 1974. (abstract)
686. Shendrikar, A. D., and P. W. West. Determination of selenium in the smoke from trash burning. Environ. Lett. 5:29-35, 1973.
687. Shepherd, L., and R. E. Huber. Some chemical and biochemical properties of selenomethionine. Can. J. Biochem. 47:877-881, 1969.
688. Shortridge, E. H., P. J. O'Hara, and P. M. Marshall. Acute selenium poisoning in cattle. N.Z. Vet. J. 19:47-50, 1971.
689. Shrift, A. A selenium cycle in nature? Nature 201:1304-1305, 1964.
690. Shrift, A. Aspects of selenium metabolism in higher plants. Ann. Rev. Plant Physiol. 20:475-494, 1969.
691. Shrift, A. Biochemical interrelations between selenium and sulfur in plants and microorganisms. Fed. Proc. 20:695-702, 1961.
692. Shrift, A. Metabolism of selenium by plants and microorganisms, pp. 763-814. In D. L. Klayman and W. H. H. Günther, Eds. Organic Selenium Compounds: Their Chemistry and Biology. New York: Wiley-Interscience, 1973.
693. Shrift, A. Microbial research with selenium, pp. 241-271. In O. H. Muth, Ed. Symposium: Selenium in Biomedicine. First International Symposium, Oregon State University, 1966. Westport, Conn.: AVI Publishing Co., 1967.
694. Shrift, A. Sulfur-selenium antagonism. I. Antimetabolite action of selenate on the growth of *Chlorella vulgaris*. Amer. J. Botany 41:223-230, 1954.
695. Shrift, A. Sulfur-selenium antagonism. II. Antimetabolite action of selenomethionine on the growth of *Chlorella vulgaris*. Amer. J. Botany 41:345-352, 1954.
696. Shrift, A., and E. Kelly. Adaptation of *Escherichia coli* to selenate. Nature 195: 732-733, 1962.
697. Shrift, A., J. Nevyas, and S. Turndorf. Mass adaptation to selenomethionine in populations of *Chlorella vulgaris*. Plant Physiol. 36:502-509, 1961.
698. Shrift, A., J. Nevyas, and S. Turndorf. Stability and reversibility of adaptation to selenomethionine in *Chlorella vulgaris*. Plant Physiol. 36:509-519, 1961.
699. Shrift, A., and M. Sproul. Nature of the stable adaptation induced by selenomethionine in *Chlorella vulgaris*. Biochim. Biophys. Acta 71:332-344, 1963.
700. Shrift, A., and T. K. Virupaksha. Biosynthesis of Se-methylselenocysteine from selenite in selenium-accumulating plants. Biochim. Biophys. Acta 71:483-485, 1963.
701. Shrift, A., and T. K. Virupaksha. Seleno-amino acids in selenium-accumulating plants. Biochim. Biophys. Acta 100:65-75, 1965.
702. Shum, A. C., and J. C. Murphy. Effects of selenium compounds on formate metabolism and coincidence of selenium-75 incorporation and formic dehydrogenase activity in cell-free preparations of *Escherichia coli*. J. Bacteriol. 110:447-449, 1972.
703. Sidi, E., and Mme Bourgeois-Spinasse. Causes actuelles les plus fréquentes des alopécies féminines. Presse Med. 66:1767-1769, 1958.

704. Sillen, L. G., and A. E. Martell. Stability Constants of Metal-Ion Complexes. Special Publication No. 17. (2nd ed.) London: The Chemical Society, 1964. 754 pp.
705. Simpson, B. H. An epidemiological study of carcinoma of the small intestine of New Zealand sheep. N.Z. Vet. J. 20:91–97, 1972.
706. Sivjakov, K. I., and H. A. Braun. The treatment of acute selenium, cadmium, and tungsten intoxication in rats with calcium disodium ethylenediaminetetraacetate. Toxicol. Appl. Pharmacol. 1:602–608, 1959.
707. Slepyan, A. H. Selenium disulfide suspension in treatment of seborrheic dermatitis of the scalp. A.M.A. Arch. Derm. Syphil. 65:228–229, 1952.
708. Slinger, W. N., and D. M. Hubbard. Treatment of seborrheic dermatitis with a shampoo containing selenium disulfide. Arch. Derm. 64:41–48, 1951.
709. Smith, A. L. Separation of the Selenium Compounds in Seleniferous Plant Protein Hydrolysates by Paper Partition Chromatography. M. S. thesis, Brookings: South Dakota State College, 1949. 33 pp.
710. Smith, F. F. Use and limitations of selenium as an insecticide, pp. 41–45. In M. S. Anderson, H. W. Lakin, K. C. Beeson, F. F. Smith, and E. Thacker. Selenium in Agriculture. U.S. Department of Agriculture Handbook No. 200. Washington, D.C.: U.S. Government Printing Office, 1961.
711. Smith, M. I., K. W. Franke, and B. B. Westfall. The selenium problem in relation to public health. A preliminary survey to determine the possibility of selenium intoxication in the rural population living on seleniferous soil. U.S. Public Health Rep. 51:1496–1505, 1936.
712. Smith, M. I., and E. F. Stohlman. Further observations on the influence of dietary protein on the toxicity of selenium. J. Pharmacol. Exp. Ther. 70:270–278, 1940.
713. Smith, M. I., E. F. Stohlman, and R. D. Lillie. The toxicity and pathology of selenium. J. Pharmacol. Exp. Ther. 60:449–471, 1937.
714. Smith, M. I., and B. B. Westfall. Further field studies on the selenium problem in relation to public health. U.S. Public Health Rep. 52:1375–1384, 1937.
715. Smith, M. I., B. B. Westfall, and E. F. Stohlman. Studies on the fate of selenium in the organism. U.S. Public Health Rep. 53:1199–1216, 1938.
716. Smith, M. I., B. B. Westfall, and E. F. Stohlman, Jr. The elimination of selenium and its distribution in the tissues. U.S. Public Health Rep. 52:1171–1177, 1937.
717. Solymosi, F. Direct hypohalite titration of selenite in alkaline medium. Chem. Anal. 52:42–43, 1963.
718. Søndegaard, E. Selenium and vitamin E interrelationships, pp. 365–381. In O. H. Muth, Ed. Symposium: Selenium in Biomedicine. First International Symposium, Oregon State University, 1966. Westport, Conn.: AVI Publishing Co., 1967.
719. Spallholz, J. E., J. L. Martin, M. L. Gerlach, and R. H. Heinzerling. Immunologic responses of mice fed diets supplemented with selenite selenium. Proc. Soc. Exp. Biol. Med. 143:685–689, 1973.
720. Spare, C.-G., and A. I. Virtanen. On the occurrence of free selenium containing amino acids in onion (*Allium cepa*). Acta Chem. Scand. 18:280–282, 1964.
721. Spencer, R. P., and M. Blau. Intestinal transport of selenium-75 selenomethionine. Science 136:155–156, 1962.
722. Spencer, R. P., G. Montana, G. T. Scanlon, and O. R. Evans. Uptake of selenomethionine by mouse and in human lymphomas, with observations on selenite and selenate. J. Nucl. Med. 8:197–208, 1967.
723. Springer, S. E., and R. E. Huber. Sulfate and selenate uptake and transport in wild and in two selenate-tolerant strains of *Escherichia coli* K 12. Arch. Biochem. Biophys. 156:595–603, 1973.

724. Sprinker, L. H., J. R. Harr, P. M. Newberne, P. D. Whanger, and P. H. Weswig. Selenium deficiency lesions in rats fed vitamin E supplemented rations. Nutr. Rep. Int. 4:335-340, 1971.
725. Stadtman, T. C. Selenium biochemistry. Proteins containing selenium are essential components of certain bacterial and mammalian enzyme systems. Science 183: 915-922, 1974.
726. Stahl, Q. R. Preliminary Air Pollution Survey of Selenium and Its Compounds. A Literature Review. National Air Pollution Control Administration Publication No. APTD 69-47. Raleigh, N.C.: U.S. Department of Health, Education, and Welfare, 1969. 75 pp.
727. Stanford, G. W., and O. E. Olson. The effect of low concentrations of selenium upon the growth of grain. Proc. S. Dakota Acad. Sci. 19:25-31, 1939.
728. Steele, R. H., and D. L. Wilhelm. The inflammatory reaction in chemical injury. II. Vascular permeability changes and necrosis induced by intracutaneous injection of various chemicals. Brit. J. Exp. Path. 48:592-607, 1967.
729. Stoklasa, J. Über die Einwirkung des Selens auf den Bau- und Betriebs-stoffwechsel der Pflanze bei Anwesenheit der Radioaktivität der Luft und des Bodens. Biochem. Z. 130:604-643, 1922.
730. Stolzenberg, J. Uptake of ^{75}Se-selenomethionine by hepatoma. J. Nucl. Med. 13: 565-566, 1972.
731. Stone, J. R. Selenium, pp. 626-634. In C. A. Hampel, Ed. The Encyclopedia of the Chemical Elements. New York: Reinhold Book Corp., 1968.
732. Strock, L. W. The distribution of selenium in nature. Amer. J. Pharm. 107:144-157, 1935.
733. Suchkov, B. P. Selenium content in major nutrients consumed by the population of the Ukrainian SSR. Vopr. Pitan. 30(6):75-77, 1971. (in Russian)
734. Supplee, W. C. Feather abnormality in poults fed a diet deficient in vitamin E and selenium. Poultry Sci. 45:852-854, 1966.
735. Suzuki, S., S. Koizumi, H. Harada, K. Ito, and T. Totani. Hygienic chemical studies on harmful elements. II. Selenium content of food, soil, and water. Tokyo Toritsu Eisei Kenkyusho Kenkyu Nempo 22:153-158, 1970. (in Japanese)
736. Suzuki, Y., K. Nishiyama, Y. Matsuka, M. Oe, S. Kuwai, Y. Takano, T. Tajiri, M. Oshima, and M. Miyazaki. Studies on selenium poisoning. Part I. Effects of exposure to air pollution on the community health in the neighbourhood of a selenium-refining plant. Skikoku Acta Med. 14:846-854, 1959. (in Japanese)
737. Suzuoki, T. A geochemical study of selenium in volcanic exhalation and sulfur deposits. J. Chem. Soc. Jap. 37:1200-1206, 1964.
738. Swanson, V. E. Composition and trace-element content of coal and power plant ash. Part 2. In Appendix J of Southwest Energy Study. Report of the Coal Resources Work Group. U.S. Geological Survey, open file report, 1972. 61 pp.
739. Sweeny, P. R., and R. G. Brown. Ultrastructural changes in muscular dystrophy. I. Cardiac tissue of piglets deprived of vitamin E and selenium. Amer. J. Path. 68: 479-492, 1972.
740. Symanski, H. Ein Fall von Selenwasserstoffvergiftung. Dtsch. Med. Wochenschr. 75:1730, 1950.
741. Tabor, E. C., M. M. Braverman, H. E. Bumstead, A. Carotti, H. M. Donaldson, L. Dubois, and R. E. Kupel. Tentative method for analysis for selenium content of atmospheric particulate matter. Health Lab. Sci. Suppl. 7(1):96-101, 1970.
742. Tank, G., and C. A. Storvick. Effect of naturally occurring selenium and vanadium on dental caries. J. Dent. Res. 39:473-488, 1960.

743. Tappel, A. L. Free-radical lipid peroxidation damage and its inhibition by vitamin E and selenium. Fed. Proc. 24:73-78, 1965.
744. Tappel, A. L., and K. A. Caldwell. Redox properties of selenium compounds related to biochemical function, pp. 345-361. In O. H. Muth, Ed. Symposium: Selenium in Biomedicine. First International Symposium, Oregon State University, 1966. Westport, Conn.: AVI Publishing Co., 1967.
745. Taussky, H. H., A. Washington, E. Zubillaga, and A. T. Milhorat. Selenium content of fresh eggs from normal and dystrophic chickens. Nature 200:1211, 1963.
746. Taylor, F. B. Significance of trace elements in public, finished water supplies. J. Amer. Water Works Assoc. 55:619-623, 1963.
747. Thapar, N. T., E. Guenthner, C. W. Carlson, and O. E. Olson. Dietary selenium and arsenic additions to diets for chickens over a half cycle. Poultry Sci. 48:1988-1933, 1969.
748. Thomas, C. G., Jr., F. D. Pepper, and J. Owen. Differentiation of malignant from benign lesions of the thyroid gland using complementary scanning with [75]selenomethionine and radioiodide. Ann. Surg. 170:396-408, 1969.
749. Thompson, J. N., and M. L. Scott. Impaired lipid and vitamin E absorption related to atrophy of the pancreas in selenium-deficient chicks. J. Nutr. 100:797-809, 1970.
750. Thompson, J. N., and M. L. Scott. Role of selenium in the nutrition of the chick. J. Nutr. 97:335-342, 1969.
751. Thorvaldson, T., and L. R. Johnson. The selenium content of Saskatchewan wheat. Can. J. Res. 18:138-150, 1940.
752. Throop, L. J. Spectrophotometric determination of selenium in steroids. Anal. Chem. 32:1807-1809, 1960.
753. Tilton, R. C., H. B. Gunner, and W. Litsky. Physiology of selenite reduction by enterococci. I. Influence of environmental variables. Can. J. Microbiol. 13:1175-1193, 1967.
754. Tinsley, I. J., J. R. Harr, J. F. Bone, P. H. Weswig, and R. S. Yamamoto. Selenium toxicity in rats. I. Growth and longevity, pp. 141-152. In O. H. Muth, Ed. Symposium: Selenium in Biomedicine. First International Symposium, Oregon State University, 1966. Westport, Conn.: AVI Publishing Co., 1967.
755. Trapp, A. L., K. K. Keahey, D. L. Whitenack, and C. K. Whitehair. Vitamin E-selenium deficiency in swine: Differential diagnosis and nature of field problem. J. Amer. Vet. Med. Assoc. 157:289-300, 1970.
756. Trelease, S. F., and O. A. Beath. Selenium: Its Geological Occurrence and Its Biological Effects in Relation to Botany, Chemistry, Agriculture, Nutrition, and Medicine. New York: The Authors, 1949. 292 pp.
757. Trelease, S. F., and A. A. DiSomma. Selenium accumulation by corn as influenced by plant extracts. Amer. J. Botany 31:544-550, 1944.
758. Trelease, S. F., A. A. DiSomma, and A. L. Jacobs. Seleno-amino acid found in *Astragalus bisulcatus*. Science 132:618, 1960.
759. Trelease, S. F., and S. S. Greenfield. Influence of plant extracts, proteins, and amino acids on the accumulation of selenium in plants. Amer. J. Botany 31:630-638, 1944.
760. Trelease, S. F., S. S. Greenfield, and A. A. DiSomma. Absorption of selenium by corn from *Astragalus* extracts and solutions containing proteins. Science 96:234-235, 1942.
761. Trelease, S. F., and H. M. Trelease. Physiological differentiation in *Astragalus* with reference to selenium. Amer. J. Botany 26:530-535, 1939.
762. Trelease, S. F., and H. M. Trelease. Selenium as a stimulating and possibly essential element for indicator plants. Amer. J. Botany 25:372-380, 1938.

763. Trinder, N., C. D. Woodhouse, and C. P. Renton. The effect of vitamin E and selenium on the incidence of retained placentae in dairy cows. Vet. Rec. 85:550–553, 1969.
764. Tsay, D.-T., A. W. Halverson, and I. S. Palmer. Inactivity of dietary trimethylselenonium chloride against the necrogenic syndrome of the rat. Nutr. Rep. Int. 2: 203–207, 1970.
765. Turina, B. Vergleichende Versuche über die Einwirkung der Selen-, Schwefel- und Tellursalze auf de Pflanzen. Biochem. Z. 129:507–533, 1922.
766. Turner, D. C., and T. C. Stadtman. Purification of protein components of the chlostridial gylcine reductase system and characterization of protein A as a selenoprotein. Arch. Biochem. Biophys. 154:366–381, 1973.
767. Tuve, T., and H. H. Williams. Metabolism of selenium by *Escherichia coli*. Biosynthesis of selenomethionine. J. Biol. Chem. 236:597–601, 1961.
768. Ulrich, J. M., and A. Shrift. Selenium absorption by excised *Astragalus* roots. Plant Physiol. 43:14–20, 1968.
769. Underwood, E. J. Trace Elements in Human and Animal Nutrition. (3rd ed.) New York: Academic Press, 1971. 543 pp.
770. U.S. Bureau of Mines. Control of Sulfur Oxide Emissions in Copper, Lead and Zinc Smelting. U.S. Bureau of Mines Information Circular No. 8527. Washington, D.C.: U.S. Department of the Interior, 1971. 62 pp.
771. U.S. Bureau of Mines. Copper, pp. 451–481. In Minerals Yearbook, 1969. Vol. 1. Metals, Minerals, and Fuels. Washington, D.C.: U.S. Department of the Interior, 1971.
772. U.S. Bureau of Mines. Copper, pp. 467–500. In Minerals Yearbook, 1970. Vol. 1. Metals, Minerals, and Fuels. Washington, D.C.: U.S. Department of the Interior, 1972.
773. U.S. Bureau of Mines. Copper, pp. 461–494. In Minerals Yearbook, 1971. Vol. 1. Metals, Minerals, and Fuels. Washington, D.C.: U.S. Department of the Interior, 1973.
774. U.S. Bureau of Mines. Copper, pp. 473–509. In Minerals Yearbook, 1972. Vol. 1. Metals, Minerals, and Fuels. Washington, D.C.: U.S. Department of the Interior, 1974.
775. U.S. Bureau of Mines. Copper, pp. 459–498. In Minerals Yearbook, 1973. Vol. 1. Metals, Minerals, and Fuels. Washington, D.C.: U.S. Department of the Interior, 1975.
776. U.S. Bureau of Mines. Selenium, pp. 1183–1184. Section in "Minor Metals" chapter in Minerals Yearbook, 1969. Vol. 1. Metals, Minerals, and Fuels. Washington, D.C.: U.S. Department of the Interior, 1971.
777. U.S. Bureau of Mines. Selenium, pp. 1223–1224. Section in "Minor Metals" chapter in Minerals Yearbook, 1970. Vol. 1. Metals, Minerals, and Fuels. Washington, D.C.: U.S. Department of the Interior, 1972.
778. U.S. Bureau of Mines. Selenium, pp. 1289–1290. Section in "Minor Metals" chapter in Minerals Yearbook, 1971. Vol. 1. Metals, Minerals, and Fuels. Washington, D.C.: U.S. Department of the Interior, 1973.
779. U.S. Bureau of Mines. Selenium, pp. 1354–1356. Section in "Minor Metals" chapter in Minerals Yearbook, 1972. Vol. 1. Metals, Minerals, and Fuels. Washington, D.C.: U.S. Department of the Interior, 1974.
780. U.S. Bureau of Mines. Selenium, pp. 1366–1369. Section in "Minor Metals" chapter in Minerals Yearbook, 1973. Vol. 1. Metals, Minerals, and Fuels. Washington, D.C.: U.S. Department of the Interior, 1975.
781. U.S. Department of Health, Education, and Welfare, Food and Drug Administration,

References

Bureau of Veterinary Medicine. Final Environmental Impact Statement: Rule Making on Selenium in Animal Feeds. Washington, D.C.: U.S. Department of Health, Education, and Welfare, 1973. 131 pp.

782. U.S. Department of Health, Education, and Welfare, Food and Drug Administration. Selenium in animal feed. Fed. Regist. 39:1355-1358, 1974.

783. U.S. Department of Health, Education, and Welfare, Food and Drug Administration. Selenium in animal feed. Proposed food additive regulation. Fed. Regist. 38: 10458-10460, 1973.

784. Usov, G. P. Effect of selenium on basic nervous processes in sheep, pp. 166-168. In M. Kruming, Ed. Materialy Respublikanskoi Konferentsii po Probleme "Mikroelementy Meditsinei Zhivotnovodstve," 1st, 1968. Baku, USSR: Izdatelstvo "ELM," 1969. (in Russian)

785. Van Vleet, J. F., W. Carlton, and H. J. Olander. Hepatosis dietetica and mulberry heart disease associated with selenium deficiency in Indiana swine. J. Amer. Vet. Med. Assoc. 157:1208-1219, 1970.

786. Van Vleet, J. F., K. B. Meyer, and H. J. Olander. Control of selenium-vitamin E deficiency in growing swine by parenteral administration of selenium-vitamin E preparations to baby pigs or to pregnant sows and their baby pigs. J. Amer. Vet. Med. Assoc. 163:452-456, 1973.

787. Vesce, C. A. Interpretation of some symptoms in selenium poisoning. Rass. Med. Ind. 17:140-143, 1948.

788. Veveris, O., S. H. Mihelson, Z. Pelekis, and I. Taure. Determination of selenium in biological materials by neutron activation analysis using 8-mercaptoquinoline. Latv. PSR Zinat. Akad. Vestis Fiz. Teh. Zinat. Ser. 2:25-28, 1969. (in Russian)

789. Virupaksha, T. K., and A. Shrift. Biochemical differences between selenium accumulator and non-accumulator *Astragalus* species. Biochim. Biophys. Acta 107: 69-80, 1965.

790. Virupaksha, T. K., and A. Shrift. Biosynthesis of selenocystathionine from selenate in *Stanleya pinnata*. Biochim. Biophys. Acta 74:791-793, 1963.

791. Virupaksha, T. K., A. Shrift, and H. Tarver. Metabolism of selenomethionine in selenium accumulator and non-accumulator *Astragalus* species. Biochim. Biophys. Acta 130:45-55, 1966.

792. Volgarev, M. N., and L. A. Tscherkes. Further studies in tissue changes associated with sodium selenate, pp. 179-184. In O. H. Muth, Ed. Symposium: Selenium in Biomedicine. First International Symposium, Oregon State University, 1966. Westport, Conn.: AVI Publishing Co., 1967.

793. von Lehmden, D. J., R. H. Jungers, and R. E. Lee, Jr. Determination of trace elements in coal, fly ash, fuel oil, and gasoline—A preliminary comparison of selected analytical techniques. Anal. Chem. 46:239-245, 1974.

794. Wahlstrom, R. C., L. D. Kamstra, and O. E. Olson. Preventing Selenium Poisoning in Growing and Fattening Pigs. South Dakota Agricultural Experiment Station Bulletin No. 456. Brookings, South Dakota State College, Agricultural Experiment Station, 1956. 15 pp.

795. Wahlstrom, R. C., and O. E. Olson. The effect of selenium on reproduction in swine. J. Animal Sci. 18:141-145, 1959.

796. Wahlstrom, R. C., and O. E. Olson. The relation of pre-natal and pre-weaning treatment to the effect of arsanilic acid on selenium poisoning in weanling pigs. J. Animal Sci. 18:578-582, 1959.

797. Walker, G. W. R., and K. P. Ting. Effect of selenium on recombination in barley. Can. J. Genet. Cytol. 9:314-320, 1967.

798. Walsh, T., and G. A. Fleming. Selenium levels in rocks, soils and herbage from a high

selenium locality in Ireland, pp. 178-185. In International Society of Soil Science. Joint Meeting of Commission II (Soil Chemistry) and Commission IV (Soil Fertility and Plant Nutrition). Dublin, July, 1952. Transactions. Vol. 2.
799. Walter, R., D. H. Schlesinger, and I. L. Schwartz. Chromatographic separation of isologous sulfur- and selenium-containing amino acids: Reductive scission of the selenium-selenium bond by mercaptans and selenols. Anal. Biochem. 27:231-243, 1969.
800. Wands, R. C. Review by the Advisory Center on Toxicology, National Research Council, of draft report by Q. R. Stahl, Preliminary Air Pollution Survey of Selenium and its Compounds, 1969.
801. Wastell, M. E., R. C. Ewan, M. W. Vorhies, and V. C. Speer. Vitamin E and selenium for growing and finishing pigs. J. Animal Sci. 34:969-973, 1972.
802. Watkinson, J. H. Analytical methods for selenium in biological material, pp. 97-117. In O. H. Muth, Ed. Symposium: Selenium in Biomedicine. First International Symposium, Oregon State University, 1966. Westport, Conn.: AVI Publishing Co., 1967.
803. Watkinson, J. H. Fluorometric determination of selenium in biological material with 2,3-diaminonaphthalene. Anal. Chem. 38:92-97, 1966.
804. Watkinson, J. H. Fluorometric determination of traces of selenium. Anal. Chem. 32:981-983, 1960.
805. Watkinson, J. H., and E. B. Davies. Uptake of native and applied selenium by pasture species. III. Uptake of selenium from various carriers. N.Z. J. Agric. Res. 10: 116-121, 1967.
806. Watkinson, J. H., and E. B. Davies. Uptake of native and applied selenium by pasture species. IV. Relative uptake through foliage and roots by white clover and browntop. Distribution of selenium in white clover. N.Z. J. Agric. Res. 10:122-133, 1967.
807. Wedderburn, J. E. Selenium and cancer. N.Z. Vet. J. 20:56, 1972.
808. Webb, J. S., and W. J. Atkinson. Regional geochemical reconnaissance applied to some agricultural problems in Co. Limerick, Eire. Nature 208:1056-1059, 1965.
809. Weisberger, A. S., and L. G. Suhrland. The effect of selenium cystine on leukemia. Blood 11:19, 1956.
810. Weiss, H. V., M. Koide, and E. D. Goldberg. Selenium and sulfur in a Greenland ice sheet: Relation to fossil fuel consumption. Science 172:261-263, 1971.
811. Weiss, K. F., J. C. Ayres, and A. A. Kraft. Inhibitory action of selenite on *Escherichia coli, Proteus vulgaris,* and *Salmonella thompson.* J. Bacteriol. 90:857-862, 1965.
812. West, P. W., and C. Cimerman. Microdetermination of selenium with 3,3'-diaminobenzidine by the ring oven technique and its application to air pollution studies. Anal. Chem. 36:2013-2016, 1964.
813. West, P. W., and T. V. Ramakrishna. A catalytic method for determining traces of selenium. Anal. Chem. 40:966-968, 1968.
814. Westfall, B. B., and M. I. Smith. Chronic selenosis. IV. Selenium in the hair as an index of the extent of its deposition in the tissues in chronic poisoning. Nat. Inst. Health Bull. 174:45-49, 1940.
815. Westfall, B. B., E. F. Stohlman, and M. I. Smith. The placental transmission of selenium. J. Pharmacol. Exp. Ther. 64:55-57, 1938.
816. Weswig, P. H., S. A. Roffler, M. A. Arnold, O. H. Muth, and J. E. Oldfield. *In vitro* uptake of selenium-75 by blood from ewes and their lambs on different selenium regimens. Amer. J. Vet. Res. 27:128-131, 1966.
817. Whanger, P. D., O. H. Muth, J. E. Oldfield, and P. H. Weswig. Influence of sulfur on incidence of white muscle disease in lambs. J. Nutr. 97:553-562, 1969.

References

818. Whanger, P. D., N. D. Pederson, D. H. Elliot, P. H. Weswig, and O. H. Muth. Free plasma amino acids in selenium-deficient lambs and rats. J. Nutr. 102:435-442, 1972.
819. Whanger, P. D., N. D. Pedersen, and P. H. Weswig. Selenium proteins in ovine tissues. II. Spectral properties of a 10,000 molecular weight selenium protein. Biochem. Biophys. Res. Commun. 53:1031-1035, 1973.
820. Whanger, P. D., and P. H. Weswig. Selenium and vitamin E metabolism in the rat. J. Anim. Sci. 31:213, 1970.
821. Whanger, P. D., P. H. Weswig, O. H. Muth, and J. E. Oldfield. Selenium and white muscle disease: Effect of sulfate and energy levels on plasma enzymes and ruminal microbes. Amer. J. Vet. Res. 31:965-972, 1970.
822. Whanger, P. D., P. H. Weswig, O. H. Muth, and J. E. Oldfield. Tissue lactic dehydrogenase, glutamic-oxalacetic transaminase, and peroxidase changes of selenium-deficient myopathic lambs. J. Nutr. 99:331-337, 1969.
823. Whanger, P. D., P. H. Weswig, O. H. Muth, and J. E. Oldfield. Tissue lysosomal enzyme changes in selenium-deficient myopathic lambs. J. Nutr. 100:773-780, 1970.
824. Whitehead, E. I., C. M. Hendrick, and F. M. Moyer. Studies with selenium 75. II. Comparison of selenium and sulfur metabolism in wheat. Proc. S. Dakota Acad. Sci. 34:52-57, 1955.
825. Williams, K. T., and H. G. Byers. Occurrence of selenium in pyrites. Ind. Eng. Chem. Anal. Ed. 6:296-297, 1934.
826. Williams, K. T., and H. G. Byers. Occurrence of selenium in the Colorado River and some of its tributaries. Ind. Eng. Chem. Anal. Ed. 7:431-432, 1935.
827. Williams, K. T., and H. G. Byers. Selenium compounds in soils. Ind. Eng. Chem. 28:912-914, 1936.
828. Williams, K. T., and H. G. Byers. Selenium in deep sea deposits. Ind. Eng. Chem. News Ed. 13:353, 1935.
829. Williams, K. T., and H. W. Lakin. Determination of selenium in organic matter. Ind. Eng. Chem. Anal. Ed. 7:409-410, 1935.
830. Williams, K. T., H. W. Lakin, and H. G. Byers. Selenium Occurrence in Certain Soils in the United States with a Discussion of Related Topics. Fourth Report. U.S. Department of Agriculture Technical Bulletin No. 702. Washington, D.C.: U.S. Department of Agriculture, 1940. 59 pp.
831. Williams, K. T., H. W. Lakin, and H. G. Byers. Selenium Occurrence in Certain Soils in the United States with a Discussion of Related Topics. Fifth Report. U.S. Department of Agriculture Technical Bulletin No. 758. Washington, D.C.: U.S. Department of Agriculture, 1941. 69 pp.
832. Witting, L. A., and M. K. Horwitt. Effects of dietary selenium, methionine, fat level and tocopherol on rat growth. J. Nutr. 84:351-360, 1964.
833. Woolfolk, C. A., and H. R. Whiteley. Reduction of inorganic compounds with molecular hydrogen by *Micrococcus lactilyticus*. I. Stoichiometry with compounds of arsenic, selenium, tellurium, transition and other compounds. J. Bacteriol. 84:647-658, 1962.
834. Wright, P. L. The absorption and tissue distribution of selenium in depleted animals, pp. 313-328. In O. H. Muth, Ed. Symposium: Selenium in Biomedicine. First International Symposium, Oregon State University, 1966. Westport, Conn.: AVI Publishing Co., 1967.
835. Wright, P. L., and M. C. Bell. Comparative metabolism of selenium and tellurium in sheep and swine. Amer. J. Physiol. 211:6-10, 1966.
836. Wright, P. L., and M. C. Bell. Selenium and vitamin E influence upon the in vitro uptake of Se^{75} by ovine blood cells. Proc. Soc. Exp. Biol. Med. 114:379-382, 1963.

837. Wright, P. L., and M. C. Bell. Selenium-75 metabolism in the gestating ewe and fetal lamb: Effect of dietary α-tocopherol and selenium. J. Nutr. 84:49–57, 1964.
838. Wright, P. L., and F. R. Mraz. Influence of dietary selenium and age upon the metabolism of selenium-75. Poultry Sci. 43:947–954, 1964.
839. Wu, M., and J. T. Wachsman. Effect of selenomethionine on growth of *Escherichia coli* and *Bacillus megaterium*. J. Bacteriol. 104:1393–1396, 1970.
840. Wu, S. H., J. E. Oldfield, O. H. Muth, P. D. Whanger, and P. H. Weswig. Effect of selenium in reproduction. Proc. Ann. Meet. W. Sect. Amer. Soc. Animal Sci. 20:85–89, 1969.
841. Wu, S. H., J. E. Oldfield, P. D. Whanger, and P. H. Weswig. Effect of selenium, vitamin E and antioxidants on testicular function in rats. Biol. Reprod. 8:625–629, 1973.
842. Yamamoto, L. A., and I. H. Segel. The inorganic sulfate transport system of *Penicillium chrysogenum*. Arch. Biochem. Biophys. 114:523–538, 1966.
843. Zingaro, R. A., and W. C. Cooper. Selenium. New York: Van Nostrand Reinhold Co., 1974. 835 pp.
844. Zoller, W. H., E. S. Gladney, and R. A. Duce. Atmospheric concentrations and sources of trace metals at the South Pole. Science 183:198–200, 1974.
845. Zweifach, B. W. Functional Behavior of the Microcirculation. Springfield, Ill.: Charles C Thomas, 1961. 149 pp.

Index

Absorption of selenium
  by plants, 66–67, 68 71
  gastrointestinal, 51–52
  through the lungs, 52
  through the skin, 52
Accumulator plants, 12, 66, 69, 72
Acid, seleno-amino, 63, 64, 76–78
Adsorption of selenites, 25
Agricultural uses of selenium, 37, 141
  as foliar sprays, 39
  as pesticide, 37
  for livestock, 37–39, 40
  to control dermatitis, pruritis, and mange, 37
Air sampling, 26–27, 135–36
Alfalfa
  effect of selenium on growth of, 70
  selenium concentration, 5
Alkali disease, 70, 107, 110, 111
Analytic methods for selenium
  by separation from interfering substances, 137
  destructive, 137
  in air, 135–36
  in animal tissues, 135
  in plants, 134–35
  in water, 135
  nondestructive
    neutron activation analysis, 136–37
    x-ray fluorescence, 136
  problems with, 134
  to measure selenium, 138
Animals. *See also* Domestic animals; Laboratory animals; Livestock
  absorption of selenium by, 51–52
  bioconversion of selenium in, 60–64
  chemical changes in selenium in, 7
  distribution of selenium in, 52–56
  effect of selenium toxicity on, 104
  excretion of selenium in, 56–60
  prophylactic and therapeutic use of selenium in, 80–81
  reducing toxic effects of selenium in, 64–66
Anticarcinogenesis, selenium and, 124–25
  recommended research on, 151
Arsenic
  metabolic interaction with selenium, 65–66, 147
  to decrease selenium toxicity, 65–66, 109
  toxicity of trimethylselenonium with, 61, 66
Atomic-absorption spectrometry, to analyze selenium, 138, 146
Atomic properties of selenium, 1
Atmosphere, selenium in
  analysis of, 135–36
  cycling of, 43
  measurement of, 27

source of, 26
Australia, dietary level of selenium in, 18

Beans, selenium in, 16
Bentonite, selenium in, 11, 12
Bile, excretion of selenium in, 57, 65
Biochemistry of selenium, 6–8
Bioconversion of selenium
  by methylation, 60–61
  in tissues, 61–64
Biosynthesis, of seleno-amino acid, 76–78
Blood
  distribution of selenium through, 54–55
  selenium in, 92
Bottom ash, selenium in, 22

Cadmium
  effect on selenium retention, 98–99
  in vascular reactions involving selenium, 109
  metabolic interaction with selenium, 65, 147
Carcinogenesis
  addition of selenium to test for, 38, 39
  experiments on selenium-induced, 120–23, 145–46
  inhibitory effect of selenium on, 94–95
  inverse relation of selenium with, 119, 124–25
  selenium therapy for, 125
  selenium to test for, 125
Cardiac myopathy, 84, 85
Cardiovascular disease, selenium deficiency and, 94
Cattle. *See* Livestock
Chemical properties, 1, 139–40
  of elemental selenium, 4–5
  of selenate–selenium, 2
  of selenide–selenium, 5–6
  of selenite–selenium, 4
Chromatographic techniques
  gas, 138
  ion exchange, 62
  paper, 61–62, 71
Chrome plating, selenium used in, 35
Coal
  emission of selenium from burning of, 47
  selenium in, 21, 22–23
Coffee, selenium in, 18
Concentration of selenium
  dietary differences in, by country, 18–20
  in atmosphere, 26–27
  in earth's crust, 9, 140
  in feces, 5
  in feedstuffs, 13, 40
  in foods, 13–30
  in hair, 58
  in internal organs, 52–54
  in limestone, 10
  in plants, 2, 5, 67
  in pyrites, 5–6
  in rocks, 2, 9, 10–11
  in sandstone, 10
  in soil, 2, 9, 11, 12, 13
  in sulfide ores, 5–6
  in sulfur, 10
  in testes, 96, 97, 98, 99
  in urine, 6
  in water, 23–25, 140
  nutrition and, 92–94
Consumption of selenium
  by agriculture
    for control of dermatitis, pruritis, and mange in dogs, 37
    for pesticides, 37
    in foliar sprays, 39
    in soil applications, 39
    to prevent livestock diseases, 37–38, 40
  by industry
    as degasifier in stainless steel, 35
    for electronic applications, 34–35
    for glass and allied industries, 33, 35
    in chrome plating, 35
    in pharmaceuticals preparation, 36
    in pigment manufacture, 36
    in rubber products manufacture, 36
  increase in, 33
Converter plants, 68
Copper
  production, 30
  refining
    recovery of selenium from slimes, 28
    selenium emissions from, 50
Crops. *See also* Plants
  selenium in, by U.S. regions, 14
Cycling of selenium, 41, 141
  environmental, 76
  industrial, 147
  in emissions, 44–49

natural macro-, 41-42, 147
  in atmosphere, 43
  in earth's crust, 43
  in food chain, 43-44
  in water, 44
  recommended research in, 149-50

Decopperization, 28-29
Dehydrogenase. *See* Formic dehydrogenase
Demographic studies, on relation of cancer to selenium, 124
Dental caries, relationship of selenium to, 118-19
Dermatitis
  from selenium, 117
  selenium to control, 37, 132
Diagnosis of selenosis, 118
Diagnostic scanning and labeling, selenium used in, 128
  for extracellular fluid volume, 131
  for liver, 129-30
  for neoplasms, 130
  for pancreas, 128-29
  for parathyroid and thyroid, 130
  for placental competence, 130-31
  for platelets and fibrinogen, 131
Diet. *See also* Nutrition
  levels of selenium in, by country, 18-20
  toxic levels of selenium in, 113, 114-16
Dimethyl selenide
  enzymic synthesis of, 61
  in plants, 72
  toxicity with mercury, 60
Distribution of selenium
  by blood, 54-56
  by internal organs, 52-54
  intracellular, 55-56
Domestic animals
  effect of excess selenium on reproduction of, 102-3
  nutritional myopathy in, 82-84
Drinking water
  arsenic to protect against excess selenium in, 65-66
  selenium in, 23-24

Earth's crust
  cycling of selenium in, 43
  selenium in, 9, 140
Eggs, selenium in, 14, 15, 16-17
Egypt, dietary level of selenium in, 19

Electronic industry
  selenium emissions from, 45
  use of selenium
    in photoelectric cell construction, 34, 35, 140
    in xerography, 34, 140
Electronic structure of selenium, 1
Electron transport chain, role of selenium in, 127-28
Elemental selenium
  formation of, 8
  from high-temperature decomposition of materials, 4
Emission of selenium
  from volcanos, 43
  industrial, 44
    from coal burning, 47
    from manufacturing, 45-47
    from mining, 45
    from smelting and refining, 45
    from solid-waste incinerators, 47
  pollution from, 49-50
Encephalomalacia, 88
Environmental pollution from selenium, 2, 4, 147
  control of, 50
  from industrial emissions, 49-50
  recommendations for monitoring, 149-50
Enzymes, effect of selenium on, 75-76, 77, 78, 79
Epidemiologic studies of selenium, 118, 124
Escherichia coli, selenium and, 79-80
Excess selenium
  effect on reproductive system, 102-3
  effect on vascular system, 106-9
Excretion of selenium
  by bile and pancreatic juice, 57, 65
  by feces, 56-57
  by hair, 58
  by lungs, 57, 65
  by saliva, 57-58
  by urine, 56, 64
  factors controlling, 58-60
Exposure to selenium
  effects of excess, 103
  recommended research on, 150
Extracellular fluid volume, Se sodium selenate to measure, 131
Exudative diathesis, 88, 105

Feces
  excretion of selenium by, 56–57
  selenium in animal, 5
Feedstuffs
  selenium concentration in, 13, 40
  toxic level of selenium in, 115–16
Ferric selenites, 4
Fertilizer
  increase selenium level of soil by, 2
  selenium in, 10
Fibrinogen, Se-selenomethionine to label, 131
Fish, selenium in, 17
Flavin–adenine dinucleotide, 76
Fluorometry, to analyze selenium, 138, 146
N-2-Fluorenylacetamide (FAA), 94, 121, 125
Fly ash, selenium in, 5, 22, 23
Foods
  selenium concentration in
    in coffee, 18
    in eggs, 14, 15, 16–17
    in fish, 17
    in meats, 17–18, 19
    in milk, 14, 15, 16–17, 19
    in poultry, 18
    in tea, 18
    in vegetables, 14, 15, 19
    in wheat and wheat products, 15–16
  toxicity of, 16
  selenium cycling and, 43–44
Formic dehydrogenase, 79–80
Fossil fuels, selenium in, 20
  coal, 21, 22–23
  oil, 23
  recommended research on, 149
Fungicides, selenium in, 36, 132–33

Gas chromatography, to analyze selenium, 138
Gastrointestinal tract, absorption of selenium through, 51–52
Geologic history of selenium, 11–13
Geology of selenium, 9–11
Germany, dietary level of selenium in, 19
Glass and allied industries, selenium used in, 32, 35, 140
Glutathione peroxidase, role for selenium in, 126–27, 128
Glutathione selenopersulfide, 63

Grains, selenium in, 15, 67
Grasses, selenium in, 67, 68

Hair
  selenium in, 58
  Selsun for treatment of dandruff in, 132–33
Hepatic cirrhosis, 120, 121
Hepatic necrosis, 84
Hepatitis, 120, 121
Hepatosis dietetica, from selenium deficiency, 85
Humans
  effect of excess selenium exposure on, 103
  medical uses of selenium in, 128–33, 146
  selenium toxicity in, 108–9
  selenosis in, 116–18, 145
  teratogenic effects of selenium on, 104–5
Hydrogen selenide, 5

Igneous rocks, selenium in, 9
Incinerators, selenium emissions from, 47
Indicator plants, 70, 71
Industrial uses of selenium, 4, 33–36, 140, 147
Intracellular distribution of selenium, 55–56
Ion exchange chromatography, 62
Irrigation water, selenium in, 24
Isotopes, for measurement of selenium, 1

Kidneys
  bioconversion of selenium in, 61, 63
  protection against excess selenium in, 64
Kwashiorkor, selenium to treat, 93

Laboratory animals
  effect of excess selenium on reproduction of, 103
  selenium deficiency in, 89–92
Lakes, selenium in, 25
Lambs. *See* Livestock
Limestone, selenium in, 10
Linseed meal, to protect against selenium toxicity, 66
Liver
  bioconversion of selenium in, 61, 62, 63
  protection against excess selenium in, 64

Se-selenomethionine for scanning of, 129-30
Livestock
  cardiac myopathy in, 84, 85
  hepatic necrosis in, 84
  hepatosis dietetica in, 85
  selenium deficiency in, 37-38, 82-86
  sudden death in, 84, 85
Lungs
  absorption of selenium through, 52
  excretion of selenium by, 57, 65
Lymphomas, Se-selenomethionine for scanning of, 130

Macrocycling of selenium, 41-44
Mange, selenium to control, 37
Manufacturing, selenium emissions from, 45-47
Meats, selenium in, 17-18
Medical uses of selenium
  for diagnostic scanning and labeling, 128-31, 146
  for therapeutic uses, 132-33, 146
Mercuric salts, effect on selenium retention, 101
Mercuric selenide, 5
Mercury
  metabolic interaction with selenium, 65, 147
  toxicity with dimethyl selenide, 60
Metabolic reactions of selenium
  in animals
    absorption, 51-52, 142
    bioconversion, 60-64
    distribution, 52-56
    excretion, 56-60, 142
  in domestic animals, 82-88
  interrelationships with other elements, 64-66, 73, 75, 147, 151
  in laboratory animals, 89-92
  in microorganisms
    from biosynthesis of seleno-amino acids, 76-78
    from methylation of inorganic selenium compounds, 78-79
    from reduction and oxidation of selenium compounds, 75-76
  in plants
    absorption, 66-67, 68, 71, 73-74
    chemistry of, 70-72
    distribution, 68
    reduction, 68-69, 73-74
    products of, 6
    recommended research on, 150, 151
Metallurgic uses of selenium, 35
Methionine
  selenomethionine as substitute for, 77-78
  to protect against selenium toxicity, 64
Methylation
  of inorganic selenium compounds, 78-79
  of selenium, 60-61
Micronutrient, selenium as, 70
Microorganisms, metabolism of selenium in, 74-79
Milk
  selenium in, 14, 15, 16-17
  transmission of selenium in, 100-101
Mining, selenium emissions from, 45
Monkeys, selenium deficiency in, 91-92
Mulberry heart disease, from selenium deficiency, 85
Muscular system, selenium to improve functioning of, 126
Mustard seed, selenium in, 16

Neutron activation analysis of selenium samples, 136-37, 138, 146
Nonaccumulator plants, 72
Nonheme iron proteins, role of selenium in, 127
Nutrition
  recommended research on selenium in, 151-52
  role of selenium in, 92-94, 143-44
Nutritional myopathy
  causes of, 82, 83
  clinical signs of, 83, 88
  incidence of, 83
  prevention of, 84

Oceans, selenium in, 24-25
Oil
  emission of selenium from burning of, 47
  selenium in, 23
Organic chemistry of selenium, 8
Oxidation of selenium, 2, 3, 6
  chemical weathering responsible for, 12-13
Oxidation of selenium compounds, 76

Pancreas, Se-selenomethionine for scanning of, 128–29
Pancreatic juice, excretion of selenium in, 57
Paper chromatography, to detect selenium, 61–62, 71
Parathyroid, Se-selenomethionine for scanning of, 130
Periodontal disease, selenium deficiency and, 93
Pesticides, selenium used in, 37, 119, 147
Pharmaceuticals, selenium used in preparation of, 36
Photoelectric cells, selenium used in construction of, 34–35
Physiologic role of selenium
  as antioxidant, 126, 128
  in electron transport chain, 127–28
  in glutathione metabolism, 126–27, 146
  in nonheme iron proteins, 127
  recommended research on, 152
Pigment, selenium used in manufacture of, 36
Placental competence, Se-selenomethionine to measure, 130–31
Plants
  absorption of selenium by, 66–67
  analysis of selenium in, 134–35
  chemical changes in selenium in, 7, 12, 70–72
  reduction of selenium in, 6
  selenium accumulator, 66, 69, 72
  selenium in, 2, 5, 67
  selenium uptake by, 68–69
  toxicity of selenium to, 69–70
Platelets, Se-selenomethionine to label, 131
Polarography, to analyze selenium, 138
Poultry
  effect of selenium toxicity on, 104
  selenium deficiency in, 87–88
  selenium in, 18
Power plants, selenium used by, 22
Precipitation
  of selenate, 25
  removal of selenium from atmosphere by, 25
  selenium concentration in, 27
Production of selenium
  by country, 29–31
  by volatilization, 29
  from copper refinery slimes, 28
Proteins
  role of selenium in nonheme iron, 127
  selenium in, 71
  selenium in tissue, 61–64
Pruritis, selenium to control, 37
Pyrites, selenium in, 5, 7

Radioselenate, 53, 77
Radioselenite
  depletion of selenium by injection of, 59, 62
  Escherichia coli and, 76
  metabolic reaction of, 52, 53
Radioselenium, 53
Rats, selenium deficiency in, 89–91
Recovery of selenium. *See* Production of selenium
Reducible selenium, 4, 6
Reduction
  of selenium, 68–69, 73–75
  of selenium compounds, 75–76
Refining, selenium, 45
Reproductive system
  effect of excess selenium on, 102–3
  effect of selenium deficiency on, 101–2, 144
  effect of selenium supplementation on, 105
  effect of selenosis on, 117
  female
    distribution of selenium in, 99–101
    retention of selenium in, 101
  male
    distribution of selenium in, 95–99
    retention of selenium in, 97–99
  selenium to improve functioning of, 126
  teratogenic effects of selenium on, 104–5
Retention of selenium, 58–60, 64
  administered versus dietary, 97
  by female reproductive system, 101
  by male reproductive system, 97
  effect of cadmium on, 98–99
  in tissues, 58–60
Rocks, selenium in, 2, 9, 10–11
  igneous, 9
  sedimentary, 10–11
Rubber manufacture, selenium used in, 36

Index

Saliva, excretion of selenium through, 57-58
Salmonella, tolerance to selenite, 75
Samples, selenium
  analysis of, 26-27, 136-38, 146
  collection and storage of, 26, 134-36
Sandstone, selenium in, 10
Scintigraphy, 129-30
Se-cystathionine, 71
Sedimentary rocks, selenium in, 10-11
Selenates
  added to soil, 2
  environmental pollution from, 2
  in sedimentary rock, 11
  in semiarid alkaline soils, 68
  metabolism of, 74
  properties of, 2, 140
  taken up by plant roots, 6
  toxicity of, 15
Selenites
  adsorption of, 25
  dietary, 5
  ferric, 4
  in production of formic dehydrogenase, 79
  metabolism of, 73-74
  properties of, 4, 139
  taken up by plant roots, 6
  tolerance to *Salmonella,* 75
  toxicity of, 115
Selenium
  as a micronutrient, 70
  geologic history of, 11-12
  high-purity, 29
  materials balance for, 47-49
  natural levels of, 98
Selenium compounds
  absorption of, 52
  chromatography to identify, 62
  in plants, 72-74
  in tissues, 62, 63
  methylation of inorganic, 78-79
  oxidation of, 76
  reduction of, 75-76
  toxicity of, 52
  transport antagonisms between sulfur compounds and, 74-75
  volatile, 60, 61, 71
Selenium deficiency
  effect on reproductive system, 101-2
  effect on vascular system, 105-6
  induced in animals, 80
  in laboratory animals, 89-92, 143
  in livestock, 37-38, 143
    cause of, 82
    clinical evidence of, 83, 85, 86
    treatment of, 84, 86
  in man, 92-94, 144
  in poultry, 87-88, 143
  prevention of, 81, 84
  selenious salts to control, 81, 84, 87, 93
  spermatogenesis and, 96-97
  vitamin E to control, 81, 84
Selenium poisoning. *See* Selenosis
Selenium salts
  to control selenium deficiency, 81, 84, 87
  to treat kwashiorkor, 93
Selenium sulfide, to treat tinea versicolor, 133
Selenium tolerance, 80
Selenium toxicity. *See* Toxicity of selenium
Seleno-amino acids, 63, 64, 76-78
Selenocystine, 51, 62, 63
  in male reproductive system, 97
  in plants, 72
Selenohomocystine, 74
Selenomethionine, 62, 74, 143
  as substitute for methionine, 77-78
  formation of, 77
  in gastrointestinal tract, 51-52
  in male reproductive system, 97
Selenosis
  acute, 106-7, 144
  chronic, 107, 108, 110-11, 144
  diagnosis of, 118
  effect of arsenic on, 65-66
  effect of linseed meal on, 66
  effect of sulfate on, 64
  prevention of, 116, 145
  subacute, 108
  symptoms of, 106-9, 111, 117
  treatment of, 116
  types of, 110-11
Selenotrisulfides, 6
Selsun, 37, 132
Se-methylselenocysteine, 72, 73
Se-methylselenomethionine, 72, 74
Se-selenocystathionine, 72
Se-selenomethionine, 1
  for diagnostic scanning, 129-31

Sheep. *See* Livestock
Skin, absorption of selenium through, 52
Smelting process, selenium emission from, 45
Soil
  chemical changes in selenium in, 7, 12-13
  ferric selenites in, 4
  selenates added to, 2
  selenium in, 2, 6, 11, 13, 67-68
Solubility of selenium, 2, 3, 5
Spark-source mass spectrometry, to analyze selenium, 138, 146
Spectrography, to determine selenium level, 92
Spectrometry, to analyze selenium, 138, 146
Spectrophotometry, to analyze selenium, 138
Spermatogenesis, selenium deficiency and, 96-97
Sudden infant death syndrome, selenium deficiency and, 93
Sugar, selenium in, 16
Sulfide ores, selenium in, 5, 7, 10
Sulfur
  metabolic interrelationship with selenium, 73, 75
  reduction of selenium uptake by plants by, 68
  selenium in, 10
  to counteract selenium toxicity, 64
Sulfur compounds, 73
  transport antagonisms between selenium compounds and, 74-75
Supplementation of selenium
  as additive to animal feed, 38-39, 59, 60, 85, 92
  by foliar sprays, 39
  by injection, 37-38, 90-91
  by soil application, 39
  effect on muscular system, 126
  effect on reproductive system, 105, 126
  effect on vascular system, 126
Surface water, selenium in, 23-24
Swine. *See* Livestock
Symptoms
  of excess selenium exposure, 102-3
  of selenium deficiency, 83, 84, 85, 105-6
  of selenosis, 106-9, 111, 117

Tea, selenium in, 18
Teratogenicity, role of selenium in, 104-5
Testes, selenium in, 95, 96, 97, 98, 99
Therapeutic uses of selenium
  for seborrheic dermatitis, 132-33
  for tinea versicolor, 133
Thyroid, Se-selenomethionine for scanning of, 130
Tinea versicolor, selenium sulfide to treat, 133
Tissues
  selenium concentration in
    animal, 61, 91, 135
    human, 92, 98
    plant, 67
    protein, 61-64
  selenium retention in, 58-60
α-Tocopherol. *See* Vitamin E
Tolerance to selenium, 80
Toxicity of selenium. *See also* Selenosis
  acute, 112, 113
  chronic, 113
  factors influencing, 111-13, 144-45
  from foods, 16
  in animals, 5
  in humans, 8, 108-9
  levels of concentration causing, 113, 114-16
  measurement of, 111, 113, 145
  methods for counteracting, 64-66
  recommended research on, 150-51
  to plants, 6, 69-70
Treatment
  of selenium toxicity, 64-66
  of selenosis, 116
Trimethylselenonium, 61
Tumors
  selenium-induced, 120-23
  Se-selenomethionine for scanning, 130

Ukraine, dietary level of selenium in, 18-19
United States
  copper production, 30
  dietary level of selenium, 19-20
  selenium content of coal in, 21-22
  selenium production, 30
Urine
  excretion of selenium by, 56, 64
  excretion of trimethylselenonium in, 61
  selenium in, 6, 92

Index

Vascular endothelium, selenium and, 105, 106
Vascular system
　as site for toxicity, 109
　effect of excess selenium on, 106-9
　effect of selenium deficiency on, 105-6, 144
　selenium to improve functioning of, 126
Vegetables, selenium in, 14
Venezuela, dietary level of selenium in, 19
Vitamin E
　interrelationship with selenium, 126
　selenium toxicity, and deficiency in, 115-16
　sudden infant death syndrome and, 93
　to control selenium deficiency, 81, 84, 87, 88
Volatile selenium compounds, 60, 61, 71
Volatilization, selenium
　emission from, 45
　recovery of selenium by, 29
　to reduce toxic effects of selenium, 65
Volcanos
　dispersion of selenium by, 20
　emission of selenium from, 43
　origin of selenium in, 11-12, 26

Water, selenium in
　analysis of, 135
　cycling of, 43
　importance of, 25
　in irrigation water, 24
　in lakes, 25
　in oceans, 24-25
　in sewage, 24
　in springs, 24
　in surface water, 23-24
　in wells, 24
　toxic levels of, 114-15
Weathering
　oxidation of selenium during, 12-13
　selenium dispersed by, 20
Wheat, selenium in, 15

Xerography, selenium used in, 34
X-ray emission spectrography, 92
X-ray fluorescence, 136